PROGRESS IN COLLOID & POLYMER SCIENCE

Editors: H.-G. Kilian (Ulm) and G. Lagaly (Kiel)

Volume 74 (1987)

Surface Forces and Surfactant Systems

Guest Editors: J.C. Eriksson (Stockholm), B. Lindman (Lund), and P. Stenius (Stockholm)

Springer-Verlag Berlin Heidelberg GmbH

ISBN 978-3-662-15678-0   ISBN 978-3-7985-1694-6 (eBook)
DOI 10.1007/978-3-7985-1694-6
ISSN 0340-255-X

© 1987 by Springer-Verlag Berlin Heidelberg
Originally published by Dr. Dietrich Steinkopff Verlag GmbH & Co. KG, Darmstadt in 1987
Softcover reprint of the hardcover 1st edition 1987

Chemistry editor: Heidrun Sauer; Copy editing: Deborah Marston; Production: Holger Frey.

Type-Setting: K + V Fotosatz GmbH, 6124 Beerfelden

Preface

This volume of Progress in Colloid and Polymer Science contains papers presented at the 9th Scandinavian Symposium on Surface Chemistry, which was held at The Royal Institute of Technology in Stockholm, Sweden, from June 4–6, 1986. Also included are a few papers primarily related to the EUCHEM conference "Molecular Interactions Between Surfaces" held in Saltsjöbaden, close to Stockholm, from June 1–4, 1986.

The symposium was attended by 140 scientists from Scandinavia and abroad, of which about 35 represented the industrial sector. The program included five plenary lectures, 23 oral presentations and 28 posters. The main topics of the symposium were: interaction between surfaces, adsorption of proteins, lipids and phase equilibria, micelles and microemulsions and colloidal stability. Theoretical as well as applied aspects were covered.

The session on lipids and phase equilibria was held in honour of Dr. Krister Fontell on his retirement from active duty as a senior research scientist at the Chemical Centre in Lund, Sweden.

Similarly, the session on micelles and microemulsions was held to honour Professor Ingvar Danielsson on his retirement from the chair of physical chemistry at Åbo Akademi in Finland.

Much of the current research on surfactant and lipid systems in Sweden and Finland can be traced back to the pioneering work of Professor Per Ekwall. Born in 1895, Prof. Ekwall is still active and was even a very active participant in the 9th Scandinavian Symposium. It is a great pleasure to include in this volume a recently completed paper by Per Ekwall.

Jan Christer Eriksson
Björn Lindman
Per Stenius

Contents

Krister Fontell on the occasion of his retirement

In the middle of an active scientific career, Krister Fontell has reached the respectable age of 65 years. His scientific interests have been focussed on the physical chemistry of surfactant systems notably phase diagrams, the structure of lyotropic liquid crystals, micelle formation of bile salts and reversed micellar solutions (termed microemulsions by others).

His career has been centered on three places in Scandinavia; Åbo (Turku), Stockholm and Lund. Due to the turmoil created by the Second World War, his scientific studies at Åbo Akademi had a slow start and it was not easy to pursue a scientific career in Finland in the aftermath of war. At Åbo Akademi, Krister made very important contributions to the pioneering studies of phase equilibria and aggregate structures in multicomponent surfactant systems, performed by the group led by Professor Per Ekwall. During this period, Krister spent 2 years in the USA as a research fellow at the Hormel Institute, University of Minnesota.

In 1963, Krister Fontell moved with Ekwall to establish the new Laboratory for Surface Chemistry (now Institute for Surface Chemistry) in Stockholm. He played an active and important role in building new research activities in surface and colloid chemistry and in establishing a flourishing research group.

As the next step in his career, Krister moved to Lund in 1973 to take up a position in multidisciplinary research at the Chemical Center, being actively involved in research at the Divisions of Physical Chemistry 1 and 2 and at the Division of Food Technology.

To the scientific community, Krister Fontell is known most for his careful studies of phase equilibria in three-component surfactant systems. In this area he has clearly made a long-standing contribution. In the course of the phase studies, Krister has had to confront many different experimental techniques. He is a recognized expert on X-ray diffraction studies of liquid crystalline phases. Of particular importance are his contributions to the understanding of the structures of different cubic liquid crystalline phases found in surfactant systems.

We who have had the opportunity to work in close cooperation with Krister, consider this a privilege. He has been a seemingly inexhaustible source of knowledge about surfactant systems in general. He has always provided us with background knowledge for our studies when we have asked him, and when the background knowledge was not present in the literature he has made the necessary complementary experiments. In this way, he humbly taught us the virtues of persuing scientific research in a conscientious and systematic way.

Although Krister Fontell is formally retiring from his position, his colleagues and coworkers have no reason to fear losing his contributions to science. On the contrary, he is involved in numerous research projects with scientists in Lund, elsewhere in Sweden and abroad at the same time as his own research projects are flourishing more than ever. For those of us who know Krister, it is somewhat ironic that his most important current work is highly relevant to the understanding of microemulsions.

Björn Lindman
Håkan Wennerström

Ingvar Danielsson on the occasion of his retirement

The session on micellar systems at the 9th Scandinavian Symposium on Surface Chemistry was arranged in honour of Professor Ingvar Danielsson on the occasion of his retirement from the chair of physical chemistry at Åbo Akademi, which he has occupied since 1964.

The scientific work of Ingvar Danielsson is mainly concerned with the physical chemistry of surfactant systems. It ranges from the formation of pre-micellar aggregates, micelles and solubilization in aqueous solutions to multicomponent equilibria and the properties of lyotropic liquid crystalline phases, and solutions of surfactants and water in non-polar or weakly polar solvents. Ingvar Danielsson started his scientific career in the early 1950s, when most of the fundamental work on the physical chemistry of surfactants was still concerned with their behaviour in relatively dilute aqueous solutions. Some phase equilibria, mainly of binary surfactant/water systems, were known, but scientific studies of concentrated or multicomponent systems were very few in number. Thus, Ingvar Danielsson has been actively engaged in the development of a scientific understanding of surfactant systems that includes theories of micelle formation, systematic descriptions of the phase equilibria of surfactant/water/hydrocarbon systems (lyotropic liquid crystals, reverse micellar solutions, microemulsions) and the correlation of these equilibria with technical and biological applications. In particular, he was a very active participant in the pioneering and extremely important studies of phase equilibria in multi-component systems, initiated at Åbo Akademi in the 1950s, and in studies of association equilibria and thermodynamic properties of surfactant systems during the 1960s and 1970s. Thus, we acknowledge him as one of the founders of the now internationally recognized Nordic research on surfactant systems.

As a professor, Ingvar Danielsson has been very actively engaged not only in the scientific development of his department, but also in the welfare of his pupils and co-workers. His scientific impact cannot be measured by the number and quality of his scientific papers only. Those who have worked with him recognize the importance of his influence as a source of new ideas and as an inspiring participant in many extensive scientific discussions (it is a part of Ingvar Danielsson's personality that he has always been too reticent in admitting the importance of his own impact in the new scientific developments that result from such discussions). Not less important is Ingvar Danielsson's frank and persistent defence of scientific quality as the only realistic basis for the future development of his own department and of academic education and research in general. This has preserved the standard of his department and the possibilities for his pupils to develop freely throughout decades of university reform and increasing bureaucracy. We know that carrying the responsibility for this has not been easy. We wish Ingvar Danielsson many more years of work in science, without the burden of all the tasks that we feel have often taken too heavy a toll on his time.

Per Stenius

Progress in Colloid & Polymer Science

Progr Colloid & Polymer Sci 74:3–16 (1987)

Solutions of alkali soaps and water in fatty acids
VI. Studies of the refractive index and light scattering

P. Ekwall[1], L. Mandell[2] and K. Fontell[3]

[1] Gråhundsvägen 134, Stockholm, Sweden, [2] Borgå, Finland, [3] Chemical Center, University of Lund, Sweden

Abstract: Experimental results of measurements of refractive index and light scattering yield interesting information about the structure of the L_2-phase. In accordance with previously presented studies they show that one, with respect to the role of water, has two main regions, one extending from the water-free phase up to a content of water of about 40% and the other at water contents above about 55%. An intermediate transition region lies between about 40% and 55% of water. In the first main region the incorporation of water into the acid octanoate molecules influences the magnitude of the refractive index decrement and light scattering intensity. In the latter region, the partial contributions of water and the amphiphilic substance for the magnitude of the refractive index remain almost unchanged, but here it seems to be the distribution into different domains, bordering each other, of the two types of water, that in the domains with hydrated polar groups and that in the domains with unbound bulk-water, which causes the changes in the intensity of the light scattering.

The structure of the L_2-phase, among others, the diversity on the molecular level and the very high concentrations of the active amphiphilic components in large parts of the phase, render quantitative estimations of particle size impossible on the basis of light scattering measurements. Information about phase structure obtained from other experimental investigations, however, makes it possible to show the connection between phase changes on the molecular level and changes in the light scattering intensity. This has also created a basis for some, although partially hypothetical, explanations of which factors play a role in the changes mentioned, as with information about the structure. Among the latter it may be mentioned that the amphiphilic substance, up to very high water contents, retains some of the properties that the first 40% of water have caused. The differences in properties between unbound bulk-water and water bound to the polar groups are considerable; the influence of the bulk-water on the phase properties seems to vary with the manner of its distribution and the properties of the bound water are different, depending on whether it belongs to the solvation water shell of the sodium ion, or whether it is bound *via* hydrogen bridges.

Key words: System sodium octanoate/octanoic acid/water, L_2 phase, refractive index, light scattering

Introduction

It has been shown in previous studies of the isotropic L_2-phase in the system sodium octanoate-octanoic acid water [1–5] that sodium octanoate dissolves in pure octanoic acid under the formation of an acid soap of the composition $1\,NaC_8 : 2\,HC_8$. When water is added, a hydrated acid soap of the composition $1\,NaC_8 : 3\,HC_8 : x\,H_2O$ is instead formed which, when the amount of octanoic acid is not sufficient, is transformed either into an acid soap with a decreasing content of fatty acid or into a mixture of the acid soaps $1\,NaC_8 : 3\,HC_8 : x\,H_2O$ and $1\,NaC_8 : 2\,HC_8 : x\,H_2O$ or of the latter and the acid soap $1\,NaC_8 : 1\,HC_8 : x\,H_2O$. The water is incorporated up to a fairly large content into the molecules of the acid octanoates; only at very large contents does the water occur as "unbound" bulk-water. In the non-aqueous region and at low water contents, when the molar ratio between octanoic acid and sodium octanoate is above 3, the L_2-phase constitutes a solution of acid octanoate in octanoic acid. At high water contents, the phase has the character of a solution of acid octanoate in water. Between these regions, when the molar ratio between fatty acid and neutral soap is below 3 and as long as the water is incorporated into the molecules of acid soap, there is a large region where the phase resembles a melt or a liquid substance (a fluid-hydrated acid octanoate). The shape of the particles of the acid octanoates is subject to large variations. In the octanoic acid solution their shape is more or less spherical. In the intermediate region the shape changes with increasing water content in anisometric direction to elongated rods. In the region with non-bound bulk-water the molecules of the acid soap seem to be collected in double layers that occur as coherent aggregations or as disk-shaped aggregates, respectively. The amounts of the different molecular compounds vary from zero to 100% in many regions of the phase and in the whole region with fluid-hydrated acid octanoate the amount is close to 100%.

In the preceeding paper in this series we presented results from studies of water vapour pressure and electrical conductivity [5]. We will now give an account of results obtained in the 1960s by our measurements of the refractive index and light scattering. The chemicals and methods of preparation of individual samples were the same. The studies were performed in similar series to those in the preceeding paper. The locations of the individual series are shown in the triangular phase diagram in Fig. 1. The compositions of the individual samples are given in percentage by weight.

The refractive index

The measurements of the refractive index were performed with a Zeiss dipping refractometer at the wave length of 5460 Å and at a temperature of $20° \pm 0.02 °C$. The reproducibility of the readings was about ± 3 in the fifth decimal place.

The refractive index, n, of pure, non-aqueous octanoic acid was measured to 1.42818. The value changes when increased amounts of sodium octanoate or a mixture of sodium octanoate and water are dissolved in octanoic acid (Fig. 2). For the non-aqueous series, the value increases at a constant rate up to a content of about 12% sodium octanoate; thereafter the rate slows down but again becomes constant at a somewhat lower level, between 17% and 30% (Fig. 2, curve a). An extrapolation to the concentration 36.54% gives the value for n of 1.44208, which will thus be the value of the acid sodium octanoate of the composition $1 NaC_8 : 2 HC_8$. When mixtures of water and sodium octanoate in constant ratios are dissolved in octanoic acid, the values of n increase at a slower rate than in the non-aqueous case (Fig. 2, curves b and c). In the series with molar ratio of 3.96

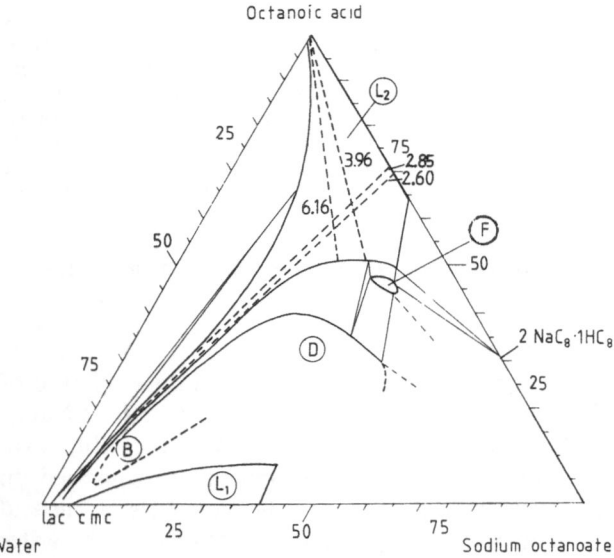

Fig. 1. Partial phase diagram for the ternary system sodium octanoate, octanoic acid and water at 20°C [1]. The locations of the individual series of samples in the study of refractive index and light scattering are marked

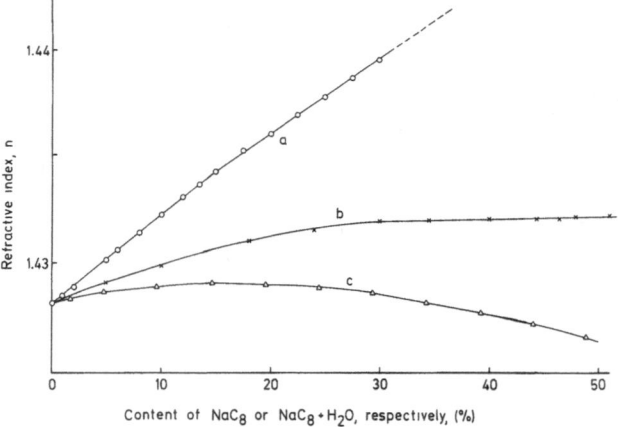

Fig. 2. The dependence of the refractive index n on the content of sodium octanoate, or a mixture of sodium octanoate and water, dissolved in octanoic acid. Series descending from the octanoic acid corner of the triangular phase diagram towards the water/sodium octanoate axis. Notations: curve (a) water-free L_2-phase; ($\bigcirc$) sodium octanoate dissolved in octanoic acid; curves (b) and (c) water containing L_2-phase; ($\times$) and ($\triangle$) mixtures of water and sodium octanoate in constant molar ratios, 3.96 and 6.16, dissolved in octanoic acid, respectively

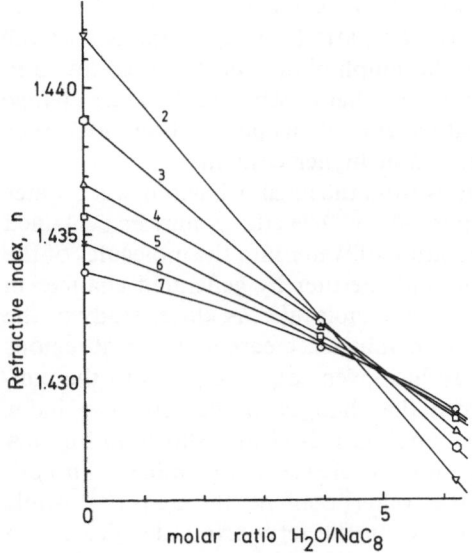

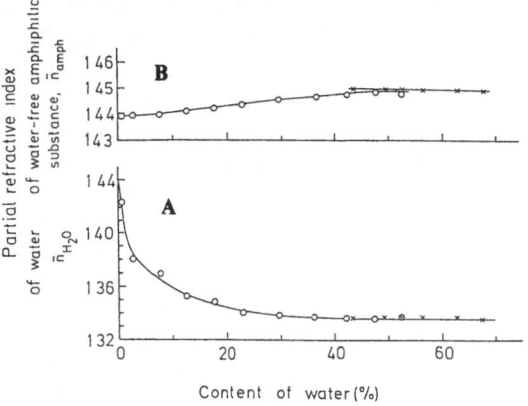

Fig. 5. The values of the partial refractive indices of water, $\bar{n}_{H_2O}$, (A), and water-free amphiphilic substance, $\bar{n}_{amph.}$, (B)

Fig. 3. The dependence of the refractive index n on the molar ratio of water to sodium octanoate for series with constant molar ratios (7, 6, 5, 4, 3 and 2) between octanoic acid and sodium octanoate

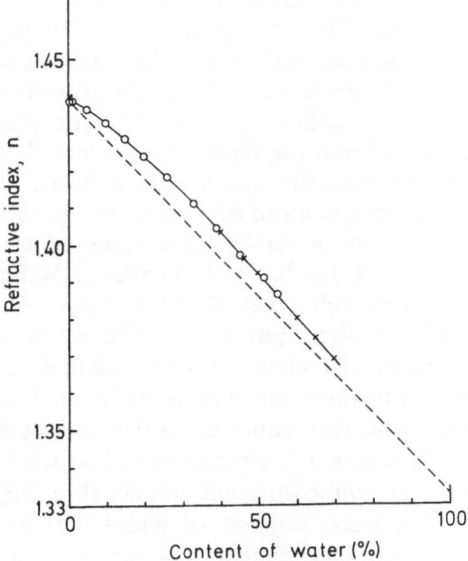

Fig. 4. The dependence of the refractive index n on the content of water in series starting from the non-aqueous octanoic acid/sodium octanoate axis, in the triangular phase diagram, and extending in the salient towards the water corner. Notations: (○) and (×) molar ratios 2.85 and 2.60, respectively, between octanoic acid and sodium octanoate. (− − −) the straight connection between the values of n for the water-free mixture of octanoic acid and sodium octanoate, and the pure water

between water and octanoate there is an increase of n up to a content of 32% of the water/octanoate mixture and thereafter the value remains almost constant. In the other series with the molar ratio 6.16, the increase in n takes place only up to 15% of the mixture, after which n begins to decrease. Under the assumption of a quantitative reaction between octanoic acid and octanoate the extrapolated values for n for the hydrated acid octanoates of the compositions $1\,NaC_8 : 3\,HC_8 : 3.96\,H_2O$ and $1\,NaC_8 : 3\,HC_8 : 6.16\,H_2O$ would be 1.43196 and 1.42755, respectively, and for the compositions $1\,NaC_8 : 2\,HC_8 : 3.96\,H_2O$ and $1\,NaC_8 : 2\,HC_8 : 6.16\,H_2O$ the values would be 1.43202 and 1.42663, respectively. An increase in water content will thus result in decreased values of the refractive index. In Fig. 3 the decreasing influence of water is illustrated; the effect is more pronounced the higher the content of acid octanoate. However, it is possible to follow this effect only up to a content of 6 moles of water per mole of sodium octanoate, that is, up to a water content of 20% at most.

Information about the effect of adding more water, up to about 55−70%, was obtained by measurements of a series that started from the non-aqueous octanoic acid/sodium octanoate axis and extended into the water-rich salient of the phase. The obtained values of n lie on a curved line above the connection line between the values for the water-free octanoic acid/sodium octanoate mixture and pure water (Fig. 4). The behaviour is thus not ideal, the values of n decrease in different rates when the water content is increased. However, from a water content of 55% up to 70% the slope of the curve, when extrapolated to 100%, gives the value 1.3336, which is rather close to that of pure water ($n = 1.33300$ at $20\,°C$). For the parent non-aqueous mixture of acid octanoate an extrapolated value of 1.4492 is obtained for *n*; this value differs considerably from the experimental value of 1.4387. The water in excess of 55% is thus incorporated into the phase with its normal value, while the amphiphilic substance up to the highest water contents continues to participate with that value of the refractive index that an addition of 40–50% of water has caused.

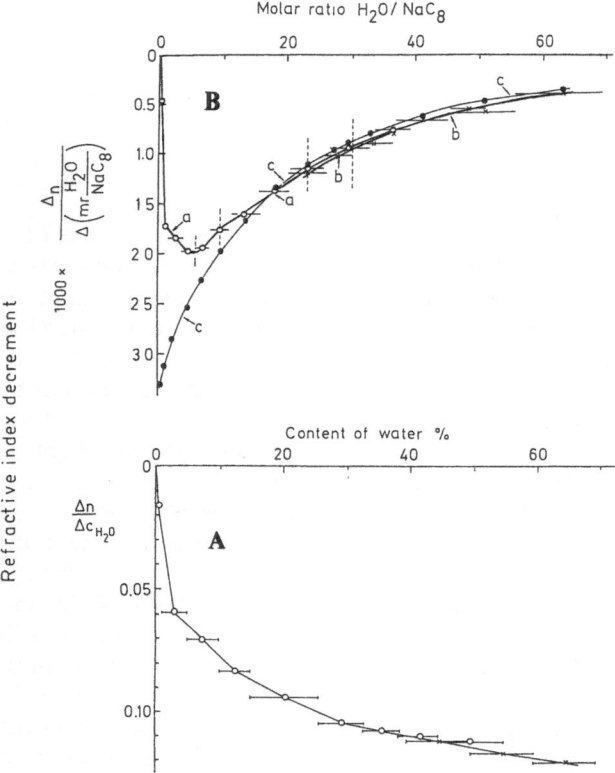

Fig. 6. The dependence of the decrement of the refractive index on the water content. (A) The decrement calculated per g of water (in g) of L_2-phase sample, $\Delta n/\Delta c_{H_2O}$, *versus* the content of water (in %). (B) The decrement calculated per mole of water per mole of sodium octanoate, $\dfrac{\Delta n}{\Delta\,mr\,H_2O/NaC_8}$ *versus* the molar ratio (*mr*) of water to sodium octanoate. Notations: (○) series with the molar ratio 2.85 (curve a), and 2.60 (×) (curve b) between octanoic acid and sodium octanoate. The solid curves through the central points of the concentration regions with constant decrement, are intended only to mark the changes in the decrement when the water content is increased. The dashed vertical lines in B mark the border between concentration regions, inside which changes in the magnitude of the decrement occur at markedly different rates. The curve (c), (●) in B gives estimated decrements of the refractive index in an ideal mixture of water and acid sodium octanoate

Figures 5 A and B show clearly the obvious differences in the values of the partial refractive indexes of the water and of the amphiphilic substance at low and high water contents. That is, while both values change considerably at water contents below about 40%, they will be unaffected at higher contents.

The deviations from the ideal behaviour when water in amounts up to 40–50% is added, may be explained by the incorporation of water into the molecules of the acid octanoate and the thereby generated changes in the structure of these molecules. A closer study of the rate at which the *n*-values decrease in different regions of the L_2-phase has given some insight into the connection between the changes in the refractive index and the water bonding. It is clearly shown in Fig. 6A that while the total decrease in the value of *n* continues with the water content through the whole region, the changes in the refractive index decrement (per gram of added water) decrease continuously. The greatest effects occur at low water contents, where all the water is bound to the acid soap; at water contents above about 40%, where the main part of the additional water is not bound, the refractive index decrement changes only a little.

The nature of this phenomenon becomes more lucid when the decrement is expressed as the decrease of *n* when one mole of water is added per mole of sodium octanoate, $\Delta n/\Delta$ (mr H_2O/NaC_8) (Fig. 6B). In the figure, the experimental values of the decrement (curves a and b) are compared with calculated values of the decrement in an ideal mixture of water and acid octanoate (curve c). From the figure it is evident that in an ideal mixture even the first small additions of water would have a pronounced decreasing effect; this would be evidenced by a rather large value of the decrement (about $3.2 \cdot 10^{-3}$). With increased water content the decrement value then would decrease continuously, rapidly at first, but then gradually at a slower rate; above water contents of about 30 mol octanoate, the values of the decrement are below 1/3 of the original ones and they continue to decrease at a very slow rate. In reality, the decrement at low water contents changes in a quite different manner (Fig. 6B, curve a). Up to a water content of 6 mol/mol octanoate, the value of the decrement increases, but only to a value (about $2 \cdot 10^{-3}$) that is much lower than that in an ideal mixture. Thereafter, the value of the decrement begins to diminish with the increase in the water content, but up to a water content of about 22–23 mol/mol octanoate this decrease clearly differs from that of an ideal solution. Even if the decrements in the real phase and in the ideal mixture thereafter also continue to be of somewhat different sizes, both curves now run in parallel, which shows that the influence of water is similar in both cases.

The discrepancy from that of an ideal mixture depends, of course, on the interaction that occurs between water and the acid octanoate molecules. The value of the decrement increases as long as the water is incorporated into the solvation shell of the sodium ion. Thereafter, the decrease seems to be somewhat different in different concentration ranges. Up to 10 mol water, the decrement continues to have a rather high value and the rate of the decrease is large; in this range, one has to expect a breaking of the hydrogen bonds between water molecules belonging to the sodium ion solvation shells and carboxylate groups of the amphiphilic molecules [4, 5]. The decrease of the decrement continues at a somewhat reduced rate up to a water content of about $23-30$ mol/mol octanoate, where the magnitude of the decrement decreases to only about the half of its maximum value (to about $1 \cdot 10^{-3}$). An increasing number of oxygen atom sites of the polar groups, where molecules of water can be bound by hydrogen bonds, may be filled up in this range. The maximum bonding capacity of the acid soap $1 NaC_8 : 3 HC_8 : x H_2O$ has been estimated as $23-24$ mol/mol octanoate [1], and all additional water should be considered incorporated into the phase as "unbound" bulk-water. In this last region, the value of the decrement is small and continues to decrease at about the same rate as in an ideal mixture.

Obviously, the changes in the magnitude of the refractive index decrement have a connection with structural changes in the molecular complexes of the acid soap caused by the incorporation of water. Since the refractive index is a property of the electron shell of the atoms and thus varies with the bonds inside the molecules, the measurements show that the incorporation of water is connected to a bonding of the water molecules to the polar groups by hydrogen bridges. They also suggest that there is a possibility for formation of water-links between different amphiphile units, which locks them in definite positions; when the links are broken, the mobility of the individual entities increases. The measurements thus support the explanation previously advanced in this series of the possible role of the incorporation of water in the packing increase inside the acid soap molecules [2, 5].

The changes in the refractive index decrement are thus mainly connected with the incorporation of water into the molecular complexes of the acid octanoate and the ensuing changes in their structure. The study suggests step-wise changes in the decrement and thus, also in the internal structure of the molecular complexes of the hydrated acid octanoates and gives in

some cases quite clear indications about the locations of the borders between the different steps. Such borders lie at 6, 10 and maybe also at $22-23$ mol water/mol octanoate. When additional water is "unbound", the changes in the decrement will be small. Even if the measurements of the refractive index clearly show that the water is no longer bound to the polar groups, at water contents above about $23-25$ mol/mol octanoate, they will not give any firm information about how much water, in reality, is bound to the polar groups of the amphiphilic compound.

Thus, it is evident that the changes in the refractive index of the L_2-phase are connected with changes inside the domains of polar groups and water. The influence of the water is different when it is bound to the polar groups or when it exists in free state. Also, the distribution of the water within the phase is influenced; as long as the water is in some manner bound to the polar groups, it will be the location of these groups inside the molecule, as well as the shape and aggregation of the molecules, that determines the distribution of the water. That continues to be the case at least up to water contents of about 40%. The unbound bulk-water, in contrast, is freely mobile and tends to form coherent domains; this effect will be more and more evident at water contents above about 55%. These differences in the distribution of the water seem to be of importance.

Light scattering

Light scattering was measured by a Sofia Photodifusogoniometer, Model 42000. The dust particles were removed by centrifuging the samples at approximately $20000 g$ in hour-glass shaped cells (Dandliker cells) floating in glycerol, a technique which proved to be more reliable and more convenient than the filtering method. The intensity readings were taken at 90° in the same cells placed in a thermostatically controlled benzene bath at $20° \pm 0.1°C$. The wave length was 5460 Å. No dissymmetry was observed in the examined specimens. Observations of the depolarization ratio were performed only for the series of non-aqueous samples. *)

---

*) Due to external circumstances it was not possible to complete the experimental study performed at the beginning of 1968; the results presented here should thus be considered only as a first orientation about the light scattering phenomenon in the present L_2-phase.

Experimental results

For pure octanoic acid, the intensity at 90° was measured as 13.5 in the arbitrary units of our apparatus. The values increased somewhat when sodium octanoate was dissolved in the non-aqueous octanoic acid, but they remained between 15 and 16 units up to a content of 30% octanoate (Fig. 7a). The formation of acid soap with the composition $1\,NaC_8:2\,HC_8$ thus affects the intensity very little, in spite of the fact that the concentration is increased to high values.

When mixtures of sodium octanoate and water in the ratios 3.96 and 6.16 mol water/mol of octanoate were dissolved in the octanoic acid, the intensity directly decreased somewhat, to about 12 units, and remained at that level at least up to an octanoate content of 17% and 21% (where the content of water was 8% and 14%, respectively). At further addition of the mixtures, the intensity began to increase slowly and the end-values 30 and 26, respectively, were obtained near the border of the phase (at 36% and 29% of sodium octanoate, respectively) (Fig. 7b, c). The formation of the hydrated acid octanoate of the composition $1\,NaC_8:3\,HC_8:x\,H_2O$ thus does not cause any increase in the intensity before the molar ratio of octanoic acid to octanoate has fallen below 4–3.5. When the content of free acid has become zero (that is, at the molar ratio of 3 between octanoic acid and octanoate) the composition of the acid soap begins to be transformed into that of an acid octanoate with successively decreasing contents of octanoic acid or into that of a mixture of the two acid soaps of the compositions $1\,NaC_8:3\,HC_8:x\,H_2O$ and $1\,NaC_8:2\,HC_8:x\,H_2O$. The ratio between the water and the total amount of amphiphilic matter in the acid soap complexes increases simultaneously. The slow increase in intensity could thus be connected either to all or to one of these changes in the composition of the phase or of the acid soap.

From the above presented measurements it is obvious that an addition of water of up to 6 mol/mol octanoate (that is up to 13–15% of water) does not affect the light scattering readings, as long as the formation of the acid soap $1\,NaC_8:3\,HC_8:x\,H_2O$ prevails.

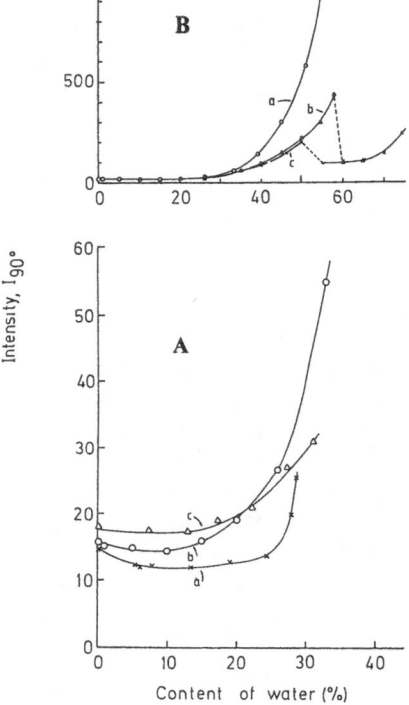

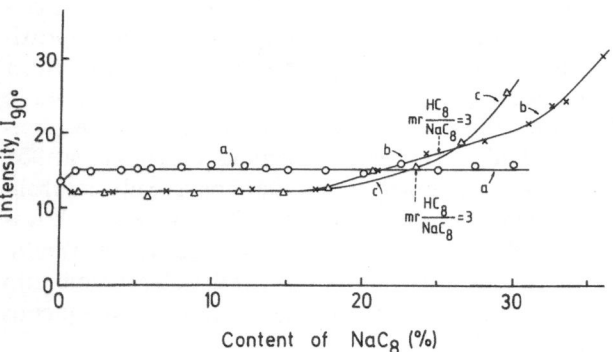

Fig. 7. The dependence of the light scattering intensity at 90° on the content of sodium octanoate in the non-aqueous series and in the series with a mixture of sodium octanoate and water. Series descending from the octanoic acid corner of the triangular phase diagram towards the water/sodium octanoate axis. Notations: curve (a) water-free L_2-phase, (○) sodium octanoate dissolved in octanoic acid; curves (b) and (c) water-containing L_2-phase; (×) and (△) mixtures of water and sodium octanoate in constant molar ratios, 3.96 and 6.16, respectively, dissolved in octanoic acid

Fig. 8. The dependence of the light scattering intensity at 90° on the content of water in series, with constant molar ratio between octanoic acid and sodium octanoate.
A. The dependence of the intensity at contents of water up to about 30%. Curve (a) (×–×) molar ratio 4.65; curve (b) (○–○) molar ratio 2.85; curve (c) (△–△) molar ratio 2.14.
B. The dependence of the intensity on the water content for series from the non-aqueous octanoic acid/sodium octanoate axis and extending in the salient towards the water corner of the phase diagram up to water contents of 50–70%. Curve (a) (○–○) molar ratio 2.85; curve (b) (△–△) molar ratio 2.69; curve (c) (×–×) molar ratio 2.60

It is evident from Fig. 8 A, however, that in series with a constant ratio between octanoic acid and sodium octanoate, a moderate increase in intensity begins at water contents above 20–25%. In the series with a constant molar ratio of 4.65 (curve a) the increase begins at about 25% of water and in the series with molar ratios of 2.85 and 2.14 an increase is quite evident at about the same water content. These observations seem to indicate that the increased intensity in light scattering is in some way associated with the increased ratio between water and amphiphile.

Measurements in series which extend from the non-aqueous octanoic acid/sodium octanoate axis into the narrow salient of the phase towards the water corner provide information about the influence of large additions of water. In Fig. 8 B results are presented for the series with the ratios 2.85, 2.69 and 2.60 mol octanoic acid/mol octanoate (curves a, b and c, respectively). Up to a water content of 10–13% there is, as already mentioned, only an extremely small influence on the intensity; thereafter, a rather slow increase begins (to the value of 55 units) that, above a water content of 33% is then followed by increases up to about 1000, 500 and 200 units, respectively, in the individual series. However, in the series with ratios 2.69 and 2.60, at water contents of between 50% and 60% the intensity thereafter abruptly falls to very low values, and increases again at a rather slow rate at further additions of water. The water content at which this break occurs seems to be shifted towards a somewhat lower value with decreasing ratios of octanoic acid to octanoate. The same holds with respect to the magnitude of the intensity peak just before the break. These observations confirm that it is not the change of the amphiphilic composition of the acid soap, as such, but the increase in the water content, that is the cause of the increase and of the ensuing break in the intensity. However, the break seems to occur at a lower total water content, the lower the molar ratio of octanoic acid to octanoate in the acid soap.

The light scattering measurements thus show that in those regions of the phase which consist of solutions of acid octanoate in octanoic acid neither the increase in the content of unhydrated acid soap $1 NaC_8 : 2 HC_8$ nor of hydrated acid soap $1 NaC_8 : 3 HC_8 : x H_2O$ causes, as such, any noticeable increase in the intensity. In that part of the water-containing region where the phase consists of fluid-hydrated acid octanoate, a connection is obvious between the intensity and the content of water. This connection continues to prevail at least as long as water is incorporated into the molecular complexes of the acid soap and bound to the polar groups [2, 3, 5].

Thus, the experimental findings themselves give some significant information about the state in the phase. Firstly, they show that a considerable change in structure occurs at water contents of between 50% and 60%. Secondly, they indicate that this change in structure is displaced towards somewhat lower contents of water when the content of octanoic acid is decreased. Thirdly, they suggest that one has a constant structure and size of particles of the formed acid soap, in those regions of the phase where it is a solution of hydrated or unhydrated acid octanoate dissolved in octanoic acid.

Correlations between changes in intensity and the phase state

The findings from the light scattering study themselves show that in the distribution of the water and water-containing parts of the molecules there is one factor that may influence the intensity. It is therefore necessary to know information obtained by other experimental studies about the division of the water into unbound bulk-water and bound hydrate-water attached to the polar groups. The distribution of the latter part of the water depends on the mutual arrangement of the molecular complexes and their structure, shape and concentration and also on the possible occurrence of an aggregation.

IR-measurements have shown that in both pure octanoic acid with its dimers and in the L_2-phase with its unhydrated or hydrated molecules of acid actanoate, it is matter of molecular complexes, inside which the individual amphiphilic entities are ordered with the polar groups towards each other and bound to each other *via* hydrogen bridges [3]. As long as this structure prevails, one should expect a basically similar influence of the molecules on the intensity of the light scattering. In the regions of the phase where one has solutions of acid octanoate in octanoic acid this seems to be the case, in spite of the existence of particles of different shapes. Octanoic acid dimers would have a shape that is somewhat elongated, while the molecules of the acid octanoates, due to their large core of polar groups, may be expected to have a more or less spherical shape. Viscosity measurements have proved this to be the case [2].

In the non-aqueous part of the L_2-phase, the particles of the acid octanoate retain their spherical shape almost up to the phase boundary; when the content of the acid octanoate has increased above about 82%, viscosity studies have revealed minor signs of a possible deviation from the spherical shape [2]. In water-containing parts of the phase the conditions are similar; here also the shape remains spherical up to a very high content of acid octanoate. However, according to the viscosity study, clear deviations from this shape occur the higher the water content of the phase [2]; in the 3.96 series this occurs at about the molar ratio of 3.8 between octanoic acid and octanoate and in the series 6.18 at the ratio 4.6.

In the water-containing L_2-phase with lower molar ratios than 3 between octanoic acid and octanoate, and more than 2 mol water/mol octanoate, no sign has been generally observed of a spherical shape of the acid octanoate particles [2]. This can be connected to, among other things, the change of the internal arrangement of the amphiphilic entities in the acid soap molecule that the incorporated water has been shown to cause and which results in a more dense packing of these entities inside the molecules.

This denser packing is clearly evident from the decreasing values of the partial molecular volume of the amphiphilic substance in the particles that are obtained when the content of water is increased [2]. In the L_2-phase with molar ratios of 2 to 3 between octanoic acid and octanoate, this decrease begins even at very low water contents and continues thereafter up to contents of 40−45%, whereupon it almost ceases. This process of change from the rather open packing, that characterizes the molecules of the non-aqueous acid octanoate 1 NaC_8 : 2 HC_8 (with a mean cross-section of about 38 $Å^2$ per hydrocarbon chain), to an extremely dense packing at the mentioned water contents has been confirmed by the IR-studies [3] which have shown that hydrocarbon chains at water contents between 42% and 60% are very close (with values for the cross-section per hydrocarbon chain that, at least locally, approach 18.5 $Å^2$). The progress of increase in packing inside the molecules is thus restricted to regions where practically all water is incorporated into the molecules of the acid octanoates and bound to their polar groups; it first ceases when sufficient amounts of additional water begin to appear as "unbound" bulk-water. The original structure of the molecules with the polar groups turning towards each other seems, however, to be retained during the mentioned process.

The incorporation of water results in an increase in the volume of the acid octanoate molecules, but this increase is bound to the central part of the molecules by its polar groups. In combination with the increased packing density of the hydrocarbon chains, this increase in volume must result in an increased anisometry of the molecules; obviously they become more elongated, rod- or ribbon-like. That means, that

they will consist of a central part, containing hydrated polar groups, from which increasingly proximate and mutually parallel hydrocarbon chains extend in two opposite directions. In that region of the phase in which this occurs, according to different observations, one has thus to consider that the original spherical shape of the acid octanoate molecules is replaced by an increasingly elongated shape with the consequences that it will have for the capability of the molecules to scatter light.

It is obvious that the light scattering intensity remains practically constant in the region where the acid soap occurs as spherical particles, irrespective of the fact that its amount increases from zero to 100%, which is compatible with the original basic structure of the molecules and their particle size remaining unchanged. In addition, the magnitude of the intensity differs only insignificantly from that of pure octanoic acid, which may indicate particles of a very similar structure and size.

Also in the region of molar ratios of between 3 and 2, where the incorporation of water has obviously resulted in the spherical shape being lost, the intensity remains uninfluenced to begin with, but this is so only up to a certain amount of water, above which one notices an increase in the intensity. This happens, however, inside the region where the volume of the molecules increases, simultaneously with their shape becoming increasingly elongated, without any change in their original structure, with the polar groups turned towards each other, and this continues to be so as long as the water is incorporated into the molecules, that is, up to a water content of 40−45%. The intensity, however, also continues to increase at somewhat higher water contents and now at an increasing rate; this faster increase continues up to a water content of 50−60%, at which a sudden decrease occurs.

Observations of another nature have shown that, as soon as the content of bulk-water in the phase has become sufficiently high, one has to consider a quite new basic type of acid octanoate molecules and aggregates [1−8]. For energetic reasons, one would expect that the polar groups here would be oriented towards the bulk-water. The diagram of the phase-equilibria (Fig. 1) shows that the two-phase zone between isotropic L_2-phase and liquid crystalline lamellar D-phase, at contents of water above 50%, contracts into an extremely narrow channel, a phenomenon that indicates that structurally both phases must be very closely related. As the D-phase consists of coherent double layers of acid octanoate with the lipophilic hydrocarbon chains facing each other and the hydrated polar groups directed outwards, towards alternating layers of bulk-water, in the L_2-phase one would also have to consider aggregations of a similar type and a distribution similar to the hydrated polar groups and the bulk-water, in relation to each other and the lipophilic parts. Measurements of vapour pressure and electrical conductivity have confirmed that the character of the phase in this region is that of a solution of acid octanoate in water and that the aggregation of the amphiphilic substance is considerable. The aggregation, however, decreases with the increase in the water content and there gradually occurs a partition into smaller aggregates without apparently altering their basic structure [5].

Thus, it seems evident that the break in the continuous increase in light scattering intensity, that occurs at a water content of 50—60%, is connected with the presence of bulk-water and with its amount having become sufficient for formation of coherent domains. The low increase in the intensity which follows, runs in parallel with the partition of the acid soap into definite aggregates.

Thus, there is in all parts of the L_2-phase a clear connection between changes in the light scattering intensity and changes in the phase structure.

Turbidity

Dimers of the octanoic acid and the acid octanoate molecules have a similar structure; this concerns not least the region with polar groups. However, the molecules of acid octanoate have the capability to incorporate large amounts of water and by this incorporation not only their volume and shape are altered but also their capability to scatter light. The values for the refractive index show that the optical polarizability per volume unit of the hydrated acid octanoates changes with their increased water content. That confirms partly that those displacements of electrons which are involved in the scattering of light are localized to the water-containing parts of the molecule, and partly that it may be the changes of character and size in just these parts that primarily influence the light scattering intensitiy. The size of these parts is not identical to the size of the kinetic units of the phase and the changes of their number may not always coincide with those of these units, which should be kept in mind.

In a system such as the present L_2-phase, with its molecular multitude, where the concentrations of the different components change from zero to 100% and their concentration in large regions is very high and the structure subject to many and radical changes, it

may be difficult to comprehend the many secondary factors that influence the scattered light intensity [9—12]. Therefore it is difficult, if not impossible, to make quantitative estimations of particle size. The available light scattering observations can be utilized only for conclusions of a more qualitative nature. We have restricted ourselves to examine whether these observations point in a quite different direction to the measurements previously obtained by other methods, or if they are compatible with them and to what extent.

The observed light scattering values have been converted into turbidity values. In this process, pure benzene was used as standard ($R_{90} = 16.3 \cdot 10^{-6}$). The obtained values are presented in Figs. 9 and 10.

In figure 9A, curve a, the turbidity values obtained for the series of sodium octanoate dissolved in pure octanoic acid are plotted against the content of the acid octanoate $1\,NaC_8 : 2\,HC_8$. The values remain practically unchanged through the region; only at the highest concentrations may a weak tendency to increase perhaps be discerned. With increased content of acid octanoate, the value of the depolarization ratio decreases rapidly from a value about unity in neat octanoic acid to a value of about 0.33 at a content of

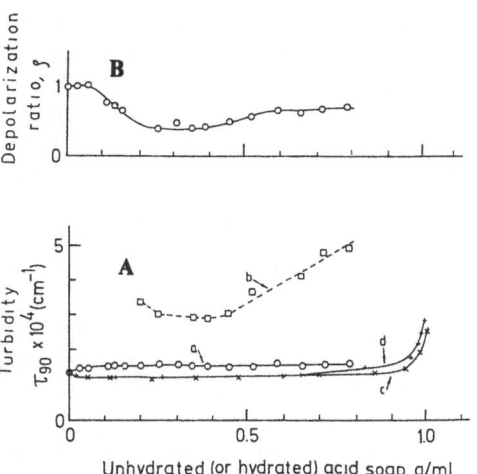

Fig. 9. Turbidity in the L_2-phase.

A. The turbidity versus the content of unhydrated, or hydrated acid octanoate, respectively, dissolved in octanoic acid. Curve (a) non-aqueous L_2-phase, uncorrected values; curve (b) non-aqueous L_2-phase, values corrected for the Cabannes-factor; curve (c) water-containing L_2-phase, a mixture of water and sodium octanoate in the molar ratio 3.96 dissolved in octanoic acid, uncorrected values. Curve (d) water-containing L_2-phase, a mixture of water and sodium octanoate in the molar ratio 6.16 dissolved in octanoic acid, uncorrected values.

B. The depolarization ratio in non-aqueous L_2-phase

about 0.5 g/ml and thereafter slowly increases up to about 0.5 (Fig. 9B). Correction in the usual manner by multiplication with the Cabannes factor results in 2−3 times higher turbidity values, and from concentrations of acid octanoate above 0.5 g/ml they show a tendency to increase (Fig. 9A, curve b).

In the two series where mixtures of sodium octanoate and water have been dissolved in octanoic acid, the content of water is so low that all water is incorporated into the molecules of the acid octanoate and therefore the turbidity may be related to the content of hydrated acid octanoate (Fig. 9A, curves c and d). In these series, the uncorrected turbidities remain practically constant up to a very high content of acid soap and that coincides with the fact that the refractive index increment is about zero at contents of hydrated acid octanoate of between 30% and 50% in the series 3.96, and between 15% and 25% in the series 6.16. In the latter series, the value of the index thereafter and up to about 45% decreases at a slow and constant rate. The value of the turbidity remains unchanged up to a molar ratio of about 3 between octanoic acid and octanoate; that is, as long as only acid octanoate $1 NaC_8 : 3 HC_8 : x H_2O$ is formed. Thereafter, one obtains a rather modest but marked increase in turbidity. Our curves suggest that this increase can hardly be the result of the insignificant increase in the total content of acid octanoate, but must be connected with the change of the amphiphilic composition; all unbound octanoic acid has been spent, and so the further addition of sodium octanoate must result in the formation of an acid soap that is poorer in fatty acid. It seems important here to try to find the reason for the increased turbidity in the distribution change between hydrated polar groups and lipophilic domains, that is a result of the transition to an acid soap which contains less fatty acid.

Our suggestion, made at an early stage, that the weak increase in intensity would indicate an incipient formation of micelles [8] is not supported by the experiences related above. In later papers of this series we will return to the question of whether there are, on the whole, possibilities for an aggregation in this part of the L_2-phase.

The samples in the series with constant molar ratios between octanoic acid and octanoate and increased amounts of water may formally be considered as solutions of water in a fluid acid soap with constant amphiphilic composition. In these cases, the turbidity has been related to the increase in the total content of water; Fig. 10 shows the dependence of the turbidity on the molar ratio between water and sodium octanoate.

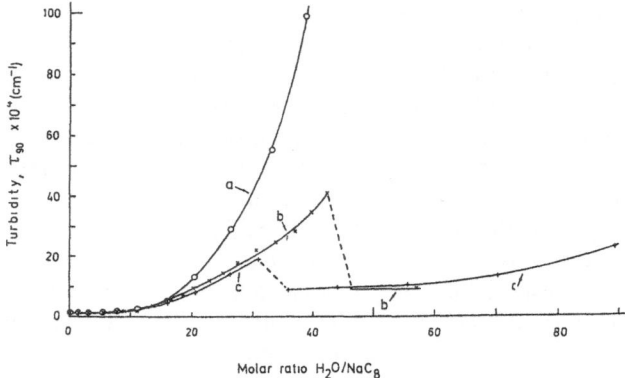

Fig. 10. Turbidity in the L_2-phase.
The turbidity for series with constant molar ratios between octanoic acid and sodium octanoate versus the molar ratio between water and sodium octanoate, uncorrected values. Curve (a) (○) $mr HC_8/NaC_8 = 2.85$; curve (b) (×) $mr HC_8/NaC_8 = 2.69$; curve (c) (+) $mr HC_8/NaC_8 = 2.60$

All the series lie in the region with molar ratios from 3.0 to 2.5 between octanoic acid and octanoate. To begin with, one has here also a long segment with almost unchanged turbidity. Up to a water content of about 6 mol water/mol octanoate (that is, about 0.15 g/ml) the turbidity remains constant; thereafter, it begins to increase very slowly, whereby the curves for the different series initially run in parallel. Firstly, at contents of water above about 16−17 mol/mol octanoate (that is about 0.34 g/ml) the increase in the turbidity becomes somewhat more rapid. Simultaneously, the curves for the different series begin to diverge; it is especially the curve for the series most rich in octanoic acid (mr = 2.85) which rapidly separates from the other two series. The differing rates at which the turbidity increases in the different series, appear in a region inside which the values for the partial refractive indexes of water and water-free amphiphilic substance are still at some contents influenced by the addition of water, but which lies close to the border where this influence terminates completely. Here, one is inside a region where, besides water being bound to the polar groups, unbound freely mobile bulk-water is added to the phase in increasing amounts. The increase in turbidity is obviously connected with this, and continues until the bulk-water has obtained such an amount that it is assembled in its own coherent domains. When that happens at contents of 35−45 mol water/mol sodium octanoate (0.5−0.6 g/ml) all influence on the values of the partial refractive indexes has ceased. The strong decrease in turbidity which then occurs thus coincides with a considerable change in the distribution between the bulk-water and the hydrated polar groups, and between them, on one side, and the lipophilic parts of the amphiphiles on the other. The reason for the same amount of water causing very different turbidities in the different series and the sudden decrease then occurring at different water contents in the different series, may be looked for in the acid octanoates. It seems to be connected with the fact that their properties are, in some respects, changed with the fatty acid content.

Concerning the deeper causes for the rapid increase in turbidity just before the sudden large decrease, and the question of possible intermediate states in the transition from a milieu free from bulk-water to one containing bulk-water, one is, for the present, confined to hypothetical explanations.

According to the view, obtained on the basis of our other experimental studies, the L_2-phase still at a water content of about $16-17$ mol/mol sodium octanoate continues to consist of fluid hydrated acid octanoates. The possibility that above this concentration local fluctuations may be the cause of the marked increase in turbidity cannot be disregarded, the more so as one is here obviously approaching a critical point concerning the structure. If it were that unbound molecules of water were initially incorporated between hydrated polar groups, then that would give rise directly to local differences of the refractive index in the phase. If that were the case, it would imply that the fluctuations and differences, respectively, would occur mainly in the fluid acid octanoate $1\,NaC_8 : 3\,HC_8 :$ H_2O complexes and would rapidly decrease with the amount of this acid soap.

However, one has also to consider another possibility viz. that the domains with water-containing parts of acid soap molecules increase. This has not been directly demonstrated, but different observations have shown that in the present region of the phase conditions really exist for a fusion of the water-containing parts of individual acid octanoate molecules. These conditions are caused by the crowding created by the incorporation of water into the molecules. The increase in the total content of water of the phase results, it is true, in a continuous decrease of the number of acid soap molecules per volume unit, but this decrease is compensated for by their molecular volume growing when water is incorporated. Thus, for a liquid, characteristic crowding is retained as long as the incorporation continues. It is known that elongated molecules or particles in liquids and concentrated solutions tend to assemble in transitional agglomerations, inside which the particles lie more or less in

parallel and this tendency increases with the length of the particles. Agglomerates of this nature in the L_2-phase must, due to the special structure of the elongated hydrated molecules of the acid octanoates, obtain a layered structure with a central layer of hydrated polar groups, which on two opposing sides are bordered with layers of hydrocarbon chains. The occurrence of such large water-containing domains could perhaps contribute to an increase in the turbidity. The stability of such a distribution would probably vary with the amphiphilic composition of the acid octanoates. The acid octanoate $1\,NaC_8 : 3\,HC_8 : x\,H_2O$ would have a greater capability to bind, screen and retain incorporated water than the compound $1\,NaC_8 : 2\,HC_8 : x\,H_2O$. The existence of such a difference has been demonstrated by measurements of electrical conductivity, concerning the capability of the acid octanoates to retain hydrated sodium ions [5].

Also, the sudden cessation of the turbidity increase would be easier to understand if agglomerates of the supposed type were formed. The agglomerates are formed only as long as the crowding prevails, and the crowding of the particles will diminish as soon as the incorporation of water into the molecules ceases and will disappear completely when the bulk-water tends to form coherent domains, which enforces a quite new distribution. The occurrence of these new domains, with bulk-water having extensions larger than the wave length of the light, would then contribute to the very low turbidity values of the phase (τ_{90} for pure water is only $0.18 \cdot 10^{-4}\,cm^{-1}$).

The supposed layered agglomerates would constitute a natural transitional state in the reorganization of the complexes of the acid soap that occurs in the transition region, and which leads to amphiphilic double layers with the hydrocarbon chains inwards, facing each other, and the polar groups outwards, towards the bulk-water. The bonding that up to now *via* the region with polar groups, has kept together the different parts of the molecules of hydrated acid octanoate, must disappear in order to facilitate the final transformation from this transitional state. That such a liberation really occurs seems to be evident from the IR-measurements [3]; they suggest that the absorption at $1900\,cm^{-1}$ completely disappears at water contents above about 42%. In the amphiphilic double layers that are the results of the transformation, a formation may instead be expected of lateral bonds between proximate molecules of acid octanoate. Also in this case, direct observations are, for the present, lacking. In later papers in this series we will return to this question and to other questions in connection with the structural transition.

The question of which of the above two hypotheses, explaining the marked increase in the turbidity, fluctuations or agglomerations, is the correct one, or if both perhaps occur simultaneously, must for the present be left unanswered. The above attempts at interpretation should be considered part of a working hypothesis which, proceeding from observations obtained from other experimental methods, attempts to elucidate which conditions on the molecular level may have a connection with the light scattering phenomenon occurring in the L_2-phase.

Discussion

The refractive index of the L_2-phase is subject to considerable changes with concentration. Concerning the magnitude of these changes, one may distinguish two different regions. At water contents below about 40% the partial refractive indexes of the water and of the water-free amphiphilic substance are changed in a characteristic manner and the changes are step-wise. They run in parallel with other changes occurring in those domains of the phase which consist of polar groups and water, and are thus localized to them; from this it is obvious that the changes are connected with the incoraportion of water into the molecules of the acid octanoates and its bonding to the polar groups. At water contents above 55% the partial index values remain more or less unchanged; the partial refractive index of the water-free amphiphilic substance here retains the value that it has obtained due to the incorporation of water in the molecules, whereas the water is incorporated into the phase with a value that agres closely with that of pure water. This is connected with the added water in this region being incorporated into the phase as unbound, freely mobile bulk-water and is assembled gradually in coherent domains. The result is that the phase water is distributed partly into domains with hydrated polar groups and partly into those with only unbound bulk-water.

The experimental data from the measurements of the light scattering already gives interesting information about the structure of the phase. The values of the intensity suggest (i) an unchanged structure inside rather large regions of the phase, (ii) a continuous change in structure inside other regions and (iii) a sudden and radical change inside a restricted region. Closer scrutiny shows a clear parallelism between changes in the intensity and changes occurring on the molecular level. The same two main regions that are evident in the measurements of the refractive index are also discerned here, from a water-free phase up to a

water content of about 40%, and at more than 55%, with an intermediate transitional region.

The turbidity remains practically unchanged as long as the phase consists of a solution of octanoic acid containing water-free acid octanoate $1\,NaC_8 : 2\,HC_8$ or hydrated acid octanoate $1\,NaC_8 : 3\,HC_8 : x\,H_2O$ with at most 6 mol water/mol sodium octanoate. When the water content increases above this value, the values for the turbidity slowly begin to increase. The same is also true in the region where the phase consists of fluid acid octanoates and has molar ratios of below 3 between octanoic acid and sodium octanoate. Here one obtains a slow increase in turbidity up to a water content of about 35 – 40%; that is, as long as the water is incorporated into the molecules of the acid octanoates; this continuous increase also occurs after reaching the transition region between 40% and 55% of water, but can reach extremely high values. At the upper limit of the latter region, a sudden and marked decrease in the turbidity occurs.

Up to a water content of about 40%, the continuous increase in the intensity may be correlated to the changes in the refractive index that are caused by the incorporation of water in the domains with polar groups. Thereafter, the changes of the partial refractive indexes are very small and some new factor begin to play a role here; this factor may have something to do with the distribution of the additional water. In the transition region, the water, initially in increasing amounts, may be incorporated between the hydrated polar groups as unbound bulk-water. The very great increase in intensity that occurs in the same region and that we have connected with the occurrence of local fluctuations, may in some way be associated with the incorporation of unbound molecules of water and with increased water domains caused by a possible formation of agglomerates. Thereafter, at contents of water between 50% and 60% the marked decrease in turbidity occurring is connected with the evident change in the distribution inside the water-containing domains of the phase, which results in the appearance of coherent regions with bulk-water. At further increase of the total water content, one observes a decreasing aggregation of the amphiphilic substance and a transition from coherent layers of hydrated acid octanoate to smaller aggregates; while the partial values of the refractive indexes remain almost unaltered, the new increase of turbidity that occurs in parallel thus also seems to be related to changes in the distribution of the lipophilic parts and the domains with hydrated polar groups of the amphiphilic substance, in relation to the increased regions with only bulk-water.

An invariance of the turbidity points to the fact that the parts of the molecules containing unhydrated or hydrated polar groups are not changed in character or in size. On the other hand, changes in the dependence of the turbidity on the water content suggest that the manner of incorporation of water is changed.

The incorporation of the molecules of water in the solvation shell of the sodium ions results in a definite, small decrease in the turbidity, while their bonding via a hydrogen bridge to the oxygen atoms of the carboxylate and carboxyl groups results in a weak increase. The turbidity will be influenced in quite a different manner when the water is not at all incorporated into the molecules of the acid soap but remains unbound; the effect, then, seem to depend on how the molecules of the free water are distributed in relation to the bound water. If molecules of unbound water are layered between the hydrated polar groups, the result seems to be a marked increase in turbidity, while one obtains a marked decrease when the free water is collected in coherent regions of bulk-water. When the amphiphilic substance with its layers of hydrated acid soap is divided into aggregates dispersed in the bulk-water, the process is connected with a new but rather slow increase in turbidity.

The picture that the study of the refractive index and the light scattering, in connection with previous experiences of the conditions in the L_2-phase has given, with respect to the structure of the phase, can be summarized as follows. The formation of unhydrated acid octanoate $1\,NaC_8 : 2\,HC_8$ results in only an insignificant increase in the intensity of the light scattering which is not influenced by an increased acid soap content. The formation of hydrated acid octanoate $1\,NaC_8 : 3\,HC_8 : x\,H_2O$ results, on the contrary, in a weak decrease in intensity, which remains unaltered as long as hydration persists in the incorporation of the water molecules belonging to the solvation shell of the sodium ion; neither is the intensity influenced by the increase in the content of this acid soap. The fact that the turbidity remains unchanged inside these regions suggests that one in them has to do with an unchanged basic structure of the molecules of the acid octanoates, and also that the size of the particles found is constant. When the amount of the water incorporated into the molecules is increased above 6 mol/mol octanoate, and the influence of water on the refractive index decrement thereby is radically changed, a slow increase in the turbidity occurs. This is the case inside the region with acid octanoate dissolved in octanoic acid, as well as in the region with fluid hydrated acid octanoate.

In the latter region, however, the rate of increase becomes more rapid at contents of water above about $17-20$ mol/mol sodium octanoate, at the same time as signs of unbound bulk-water begin to appear. The basic structure of the molecules with a central core of hydrated polar groups, however, seems to be retained as long as the increase of the turbidity is continuous. At contents of water between 50% and 60% this rapid continuous increase ceases, and the turbidity sinks abruptly to a very low value, at the same time as observations show that the amount of bulk-water has become so large that it is collected into coherent domains. This indicates that the structure of the phase undergoes a radical change here. Previous observations have indicated that in this same region the structure of molecules of the acid octanoates is changed to one with the amphiphilic units in parallel, all with the fully hydrated polar groups in the same direction and in contact with bulk-water. When the total water content is thereafter increased, a new slow increase in turbidity appears in parallel with the decrease in the pronounced aggregation of the acid octanoate decreasing and the occurrence a partition in smaller aggregates occurring.

The study of the refractive index and of the light scattering has thus contributed especially to the elucidation of the role of water in the changes in properties and structure of the L_2-phase. It has to be stressed that the task that we have taken upon ourselves is to report on our previously unpublished experimental studies of this phase from the 1960s and on the conclusions about its structure that these studies give, together with what was already known before that, based upon older studies about systems containing acid soaps. When this account has been concluded, its picture of the structure will be scrutinized with respect to the new experimental studies of this phase that have appeared during the last decade.

Acknowledgements

The work has been supported by Naturvetenskapliga Forskningsrådet (the Swedish Natural Science Research Council). The experimental measurements were performed in 1967–1968 at Ytkemiska Laboratoriet, later Ytkemiska Institutet (Laboratory for Surface Chemistry/Institute for Surface Chemistry), Stockholm, Sweden. The technical assistance of Miss B. Svensson, B. A. is gratefully acknowledged.

References

1. Ekwall P, Mandell L (1969) Kolloid-Z Z Polymere 233:938 (Part I in this series)
2. Ekwall P, Solyom P (1969) Kooloid-Z Z Polymere 233:945 (Part II in this series)
3. Friberg S, Mandell L, Ekwall P (1969) Kolloid-Z Z Polymere 233:955 (Part III in this series)
4. Lindman B, Ekwall P (1969) Kolloid-Z Z Polymere 234:1115 (Part IV in this series)
5. Ekwall P, Mandell L, Fontell K (1986) Colloid and Polymer Sci 264−542 (Part V in this series)
6. Ekwall P (1965) Wiss Z Friedrich-Schiller-Univ, Jena, Math-Naturwiss Reihe 14:181
7. Ekwall P (1967) Svensk Kemisk Tidskrift 79:605
8. Ekwall P (1969) J Colloid Interface Sci 29:16
9. Oster G (1948) Chem Rev 43:319
10. Debye P (1947) J Phys Coll Chem 51:18
11. Debye P (1949) J Phys Coll Chem 53:1
12. Huisman HF (1964) Koninkl. Nederl. Akademie van Wetenschappen, Amsterdam, Proceedings Series B 67 No 4:367, 376

Received June 20, 1984;
revised version February 24, 1987;
accepted March 1, 1987

Authors' address:

P. Ekwall
Gråhundsvägen 134
Stockholm, Sweden

Progress in Colloid & Polymer Science

Progr Colloid & Polymer Sci 74:17–30 (1987)

Some results from 50 years' research on surface forces

B. V. Derjaguin

Institute of Physical Chemistry, USSR Academy of Sciences, Moscow, USSR

Abstract: A review is presented about research on surface forces and surface interactions conducted over the past half-century, with some emphasis on the pioneering contributions of the Department of Surface Phenomena at the Institute of Physical Chemistry of the USSR Academy of Sciences.

Key words: Surface forces, interlayers, thermoosmosis, capillary and inverse osmosis, boundary phase, enthalpy density

In the early 1930s, the notion of short-range surface forces was predominant. Underlying this, were the experiments carried out by Raleigh, Perrin, and especially Langmuir. The exceptions to this were the theories of Guy-Chapman on diffuse ionic atmospheres and the calculation of molecular-surface forces on the basis of summation of London's pair-wise forces. Hardy and Hennicker may be mentioned as scientists with opposing view-points.

At that time, the author developed an essentially new approach, both experimentally and theoretically, by carrying out measurements for the first time of the interaction of surfaces separated by a liquid interlayer, as a function of its thickness [1, 2].

The concept of disjoining pressure, $\Pi(h)$, had been introduced, which was defined as equal to a difference between the thermodynamic equilibrium state pressure applied to surfaces by separating interlayer, and the pressure in the bulk phase, which the interlayer had been formed from during the process of its thinning, or with which it was in thermodynamic equilibrium [3].

The dependence of the $\Pi(h)$-isotherm of disjoining pressure, was determined by measurements carried out for an interlayer separating two solid phases [1], a solid and a gaseous phase (the wetting films [2]), and finally two gaseous phases (e.g. free films [4]).

These first experiments were accompanied both by introducing the concept of disjoining pressure into the thermodynamics of heterogeneous systems and by examining its microscopic mechanism. In the second approach, forces of different nature were considered, that could generate different components of disjoining pressure.

Firstly, dispersion (molecular) forces acting close to interfaces and electric fields connected with the formation of diffuse double layers were examined. As a rule, some component of disjoining pressure arises when the "near-to-surface" zones, in which forces are active, overlap: in the first case these are dispersion ones, and in the second, electrical ones (or those in which ionic atmospheres have been localized).

The first correct experimental determinations of the molecular component of disjoining pressure (having negative sign – that is: referring to attraction) were published by Derjaguin and Abrikosova in 1953–55 [5] for an interlayer between solids (e.g. glass, quartz, chromium, thallium halides).

These measurements had given rise to the development by Lifshitz [6] of the macroscopic theory of molecular interaction, which had begun to be extensively applied. The author and Kusakov began to measure the disjoining pressure of wetting films in 1937 [2]. These measurements had induced Derjaguin (in 1937–1940) [7] and later working with Landau (1941) [8], to develop a theory of the ionic-electrostatic component of disjoining pressure, whether for a symmetrical case (applicable to the theory of stability and coagulation of lyophobic colloids) or, somewhat later [9], for a nonsymmetrical case applicable to hetero-coagulation and flotation [10].

Simultaneously (1937), research into the disjoining pressure of wetting and adsorption-wetting layers of polar liquids [11] allowed a conclusion to be drawn about the existence of a structural component of disjoining pressure. This component was observed to arise when either two structurally-modified boundary layers overlapped, framing a liquid interlayer, or one of these was overlapped by an interface with another (gas) phase. In both cases, the liquid interlayer was found to be devoid of areas having a structure which would remain invariable as compared with that of the bulk liquid phase. It is essential that the structural component of disjoining pressure for an adsorption layer of water, on the surface of glass or fused quartz, is positive for small thicknesses (a is a branch of the whole $\Pi(h)$ isotherm; see Fig. 1), corresponding to equilibrium with the vapour pressure below the equilibrium pressure. This branch cannot be ascribed to the sum of only the molecular and the electrical component of disjoining pressure [12]. Hence, the structural component of disjoining pressure changes its sign at the point of equilibrium between the bulk phase and a plane surface. In accordance with the Derjaguin theory [13], the contact angle θ is expressed through the $\Pi(h)$ isotherm of disjoining pressure in the following manner:

$$\sigma \cos \theta = \sigma + \int_{h_0}^{\infty} \Pi(h)\,dh + \Pi(h_0)\,h_0 \qquad (1)$$

where σ is the surface tension of liquid; h_0 is the thickness of a liquid interlayer.

From this equation and from Fig. 1, it is possible to draw the following conclusion: where the structural component of disjoining pressure is substantial, both complete and incomplete wetting may be observed, depending on the isotherm of the structural component. In the latter, the contact angle depends on the

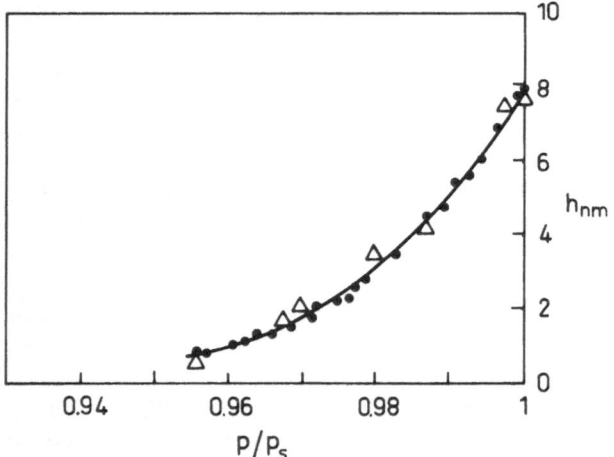

Fig. 1. Isotherm of the adsorption of water vapour on glass
$$\Pi(h) = \frac{RT}{v_m} \ln \frac{p}{p_s}$$

vapour pressure of an adjacent phase. In a remarkable work by Bangham, published in 1937 [14], the formation of very thick layers was detected during condensation of strongly supersaturated vapours of a number of polar liquids, including those of water and alcohols. Drops of the same liquids, dropped onto a platelet coated with a "supersaturated" (metastable) adsorption layer, did not coalesce with it, forming a contact angle differing from zero.

This is already sufficient to prove that the structure of a metastable condensate film is different from the "normal" structure of liquid in bulk. The disjoining pressure of a condensate being negative, its layer tends to thin our, flowing out into the regions where there is no flow of supersaturated vapour, and the condensate film can evaporate.

By judging from interference colours, Bangham evaluated the thickness of condensate films to be of the order of a micron, while be assumed the contrast range of these colours to be an indication of a changed refractive index. Consequently, the concepts of a special structure of boundary layers, including those in the metastable state, and of "structural forces" allow an adequate interpretation of Bangham's experimental results.

It seems surprising that such remarkable observations exhibiting a substantial novelty, have not been continued. As an explanation for this, their seemingly paradoxical character and the absence of theoretical approach may be mentioned. This often gives rise to biassed scepticism and a lack of attention. Generally speaking, exploration in new fields of science involves difficulties, impasses, and errors, but its genuine value and significance must still be highly estimable.

These considerations may essentially be applied to the whole concept of the structural peculiarities of boundary layers and "structural forces". Until recently, most specialists in the field of colloid-surface phenomena exhibited either a sceptical attitude towards it or a complete lack of interest, in spite of the unambiguous proofs in favour of it, which have been presented; namely, deviations of the viscosity of boundary layers from the bulk value [15–17]; birefrigence of water interlayers up to 200 Å thick, between crystalline planes in a swollen Na-montmorillonite, according to the data of Green-Kelly and Derjaguin [18]; thermo-osmosis [19]; the thermal expansion of thin water interlayers [20], and other effects. In the last few years, the situation has changed: a markedly increased attention is being paid to structural effects, so that these may be assumed to have gained general recognition. Major research stages may be considered to include the investigation of the stability of aqueous silica sols, suspensions of quartz, diamond [21]; the swelling and other properties of clays (Low [22]); direct investigations of the disjoining pressure of aqueous interlayers between quartz surfaces (cf. Rabinovich [23], Peschel and Belouschek [24], Israelachvili [25]). It is very important that Israelachvili has detected cases [25] in which water interlayers of up to 100 Å thick, between hydrophobic surfaces, exhibit a negative disjoining pressure. As has been detected by Shchukin, metastability of the same interlayers may cause the nucleation appearance of a bubble and its growth in an interlayer [26].

Wide recognition and success of the theory of Dzyaloshinsky, Lifshitz, Pitaevsky (DLP theory) concerning the molecular component of the disjoining pressure between any phases, had induced one to overlook the conditions of its rigorous application. Thus, even for solutions of nonelectrolytes in nonionic media, the condition of nonuniformity of composition of an interlayer, without which the above theory is inapplicable, may be substantially broken, owing to the formation of diffuse atmospheres of a considerable extension.

Derjaguin was the first to demonstrate this, by calculating the effective fields of forces acting on dissolved molecules close to an interface, on the basis of a particular application of the DLP theory [28]. Experiments involving an investigation of capillary osmosis, as begun in 1947 [29], enable one to substantiate and assess the diffusivity of nonionic adsorption atmospheres [30]. Simultaneously, a correction, as a first approximation, was entered into the DLP theory [31, 32]. Aside from stating corresponding conclusions, it would be appropriate to emphasize that even in the absence of ions, the disjoining pressure of solutions for a symmetrical system might become positive, thereby ensuring the stability of an interlayer [31] when either:

$$\varepsilon_1 > \varepsilon_2 \quad \text{and} \quad \frac{2}{\varepsilon_s} < \left(\frac{1}{\varepsilon_1} + \frac{1}{\varepsilon_2} \right) \tag{2}$$

or

$$\varepsilon_2 < \varepsilon_1 \quad \text{and} \quad \frac{2}{\varepsilon_s} > \left(\frac{1}{\varepsilon_1} + \frac{1}{\varepsilon_2} \right) \tag{3}$$

where ε_1 is the dielectric permittivity of the solvent, ε_2 is the permittivity of the surrounding phase. Additionally, we assume smallness of the expression:

$$\frac{A}{kTh^3} \ll 1 \tag{4}$$

where h is the thickness of the interlayer; A is the Hamaker constant; and the linearity of the dependence of the dielectric permittivity of solution, ε_s, on the concentration of solution, C. This conclusion is substantiated by a number of works [32].

This underlines the essential significance of taking into consideration the adsorption component of disjoining pressure, which is also treated in a number of other works [33–35].

Beginning with the development of the theory of stability of lyophobic colloids, in calculating the disjoining pressure, the simplest assumption is made of additivity and the mutual independence of its components. But in the work of Gorelkina and Smilga [36], Davies and Ninham [37], and Mitchell and Richmond [38], it was demonstrated that a partial screening of dispersion forces occurs in electrolyte solutions.

The mutual influence of the electrostatic and the structural component is more complex. The absence of a complete quantitative theory of structural forces interferes with the solution of the problem in a general form. The problem may be restricted, however, by assuming variation in the electrostatic component of disjoining pressure, owing to a change in the properties of the solvent (e.g. its dielectric permittivity, dissolving power) as a result of changes in its properties close to interfaces. Dukhin et al. [39] calculated the electrostatic component of disjoining pressure on the assumption that a boundary layer of thickness H, possesses a reduced dissolving power, as compared with that of the bulk of electrolyte solution.

In this case, it has to be acknowledged that reduction in the dissolving power is different for cations and anions.

The concept of the "steric repulsion" of adsorbed monolayers of surfactants (or polymers) will have to be distinguished from that of the adsorption component considered above. The former is essentially interpreted in a simplified mechanical manner. Ample literature is available on that problem, so we shall not deal with it here. We shall consider only the works in which the formation of structurally modified or diffuse ionic layers exerts a substantial influence on the corresponding repulsion forces.

Thus, Glazman [40] has proved the following: in certain cases, the formation at the lyophilic ends of adsorbed molecules, of surfactants of "solvate" boundary layers having a modified structure, is necessary to allow the setting up of a positive disjoining pressure whose radius of action exceeds the double thickness of adsorbed layers.

Dukhin and Derjaguin [4] developed a theory of the formation of a diffuse ionic atmosphere, taking into account the distribution of ions inside the adsorption layers. It has been shown that as a result, the stabilizing effect of ionic atmospheres may markedly increase.

At our Department of Surface Phenomena, quite recently and unexpectedly, a special mechanism of disjoining pressure was detected in observing mercury bubbles in hydrocarbon liquids (e.g. octane, decane, hexane, etc. [42]).

In Fig. 2 the schematic diagram of an instrument for detecting the equilibrium disjoining pressure is presented. A glass vessel (1) measuring $50 \times 20 \times 40$ mm was filled with mercury (2), onto which was poured a layer of hydrocarbon liquid (3) about 10 mm thick. Immersed in the mercury was the elbow of a bent capillary connected through a valve (6) with a medicinal syringe (7). Soldered onto the left end of the capillary was a porous glass disc (4) about 8 mm in diameter, cut from an N 3 Schott filter. Prior to carrying out an experiment, the whole capillary, including the filter and the syringe, was filled with a hydrocarbon liquid.

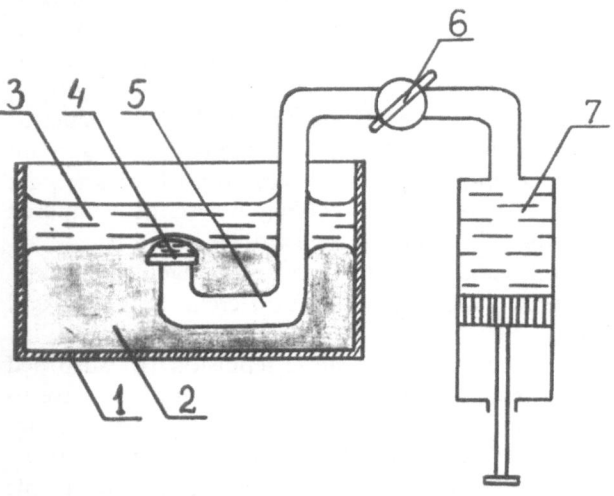

Fig. 2. Schematic diagram of the experiments on detection of the stability of mercury films (see text)

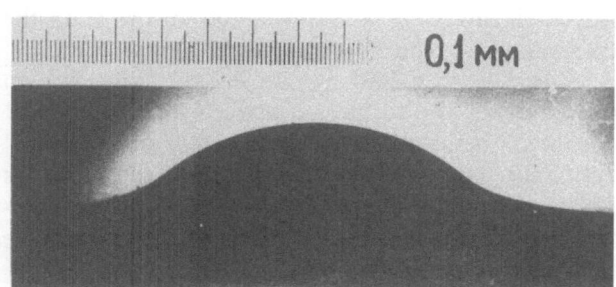

By making use of a rack, the capillary was lifted such that the upper surface of the syringe was about 0.1 to 0.3 mm below the mercury level. Then, using a micrometric device, liquid was extruded through the syringe. It formed a gradually growing convex surface (see Fig. 3). This convexity represents the external surface of a mercury film, as confined by two volumes of the hydrocarbon liquid. The highest stability of this film is observed for octane. In this case, the mercury film was preserved for an indefinite period. The life time of mercury films in other liquids is limited and varies within a wide range. All these facts are so surprising that the usual remark of sceptics may be expected: "This can't be!"

Yet the fact is that the theory developed by Derjaguin and Roldughin adequately explains this [43]. It is not possible to set forth this theory using the quantum-mechanical description of electronic gas in metal. Therefore, we will limit ourselves to a definition of its physical significance and conclusions.

As is already known, the state of gas near the surface of another condensed phase is modified in the Knudsen layer, with a thickness of the order of the molecules free path. However, this has no bearing on the thermodynamic characteristics of gas, and the overlapping of two Knudsen layers in a gas interlayer does not give rise to the setting up of a disjoining pressure.

As has been shown by Derjaguin and Martynov [44], the disjoining pressure may arise owing to the molecular component in very dense gases. As a liquid-metallic film becomes thinner, the Gibbs free energy of the system begins to vary with its thickness, when the latter becomes of the same order of magnitude as the de Broglie wavelength of at least a portion of free electrons. Simultaneously, the electronic component of the disjoining pressure of the film appears, too, being of quantum nature. Calculation gives the following formula:

$$\Pi_{el}(H) = E_f^0 n F^2 \left[\varepsilon_1 \varepsilon_2 - \frac{1}{4} \varepsilon_1^2 \right.$$
$$\left. + \frac{2}{F} \exp\left(\frac{-\gamma \cdot a}{\sqrt{E_f^0}}\right) \sin\left(2a\sqrt{E_f^0} + \varphi_0\right) \right] \quad (5)$$

where $E_f^0 = \frac{\hbar^2}{2m}(3\pi^2 n)^{2/3}$ is the Fermi energy of the free electron gas, in which $\hbar$ = the Planck constant, m = the electron mass, n = the number of free electrons per unit volume, N = the number of electrons in the film, S = the film surface area,

$$F = \left(\frac{V}{3\pi^2 N H^3}\right)^{1/3}, \quad V = H \cdot S$$

$$\varepsilon_1 = \frac{\pi}{2} - 2\left[\arctan(x_n) + x_n - x_n^2 \arctan\left(\frac{1}{x_n}\right)\right]$$

$$\varepsilon_2 = \frac{\pi}{2} - 2\arctan(x_n), \quad x_n = R_s\left(\frac{2mE_f^0}{\hbar^2}\right)^{-1/2}$$

R_s = the inverse of the characteristic distance, at which electrons penetrate the medium surrounding the film, γ = the width of the energy levels of electrons, depending on their scattering by ions, impurities, and on their interfaces with adjacent phases

$$a = H\left(\frac{2m}{h^2}\right)^{1/2}$$

$$\varphi_0 = \arccos\left[\frac{(1-x_n^2)}{(1+x_n^2)}\right].$$

From the above Eq. (5), it is apparent that the disjoining pressure consists of two cofactors: the first determines the decay which is inversely proportional to the film thickness; the second cofactor represents a sum made up of a term which is independent of the film thickness, and a term which simultaneously oscillates and decays as a function of the film thickness, H. When the film thickness corresponds to the disjoining pressure oscillation range, then, whilst taking into account the negative molecular component, the metastable state of the film is practically unstable, inasmuch as the disjoining pressure barriers are extremely narrow, of the order of 1 Å. Thicker films, when oscillations decay, may be quite stable, which explains the experiments carried out on mercury films in octane.

The stability of a mercury film which, in other hydrocarbons is limited in time, and its marked dependence on the nature of them, are explained by the dependence of the permanent term in the square brackets of Eq. (5) on the nature of the surrounding phase. This dependence is attributable to the parameter x_n, which is proportional to the value of R_s. The latter, in accordance with the developed theory, is determined by the character of collision of electrons with an adjacent phase − the work function and the probability of nonelastic repulsion (scattering).

Figure 4 illustrates a very sensitive dependence of the stability of films on the parameter x_n. When the value of x_n decreases from 1.2 to 1.15, the sum of the

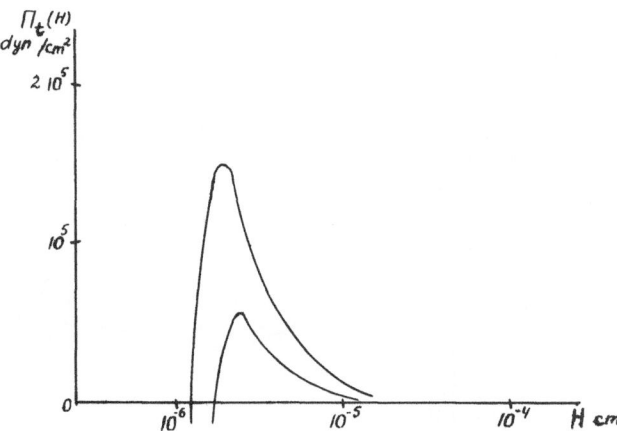

Fig. 4. Determination of the boundaries and conditions of stability of mercury films. The upper curce $x = 1.2$, the lower one $x = 1.15$

molecular and the electronic component of disjoining pressure falls off drastically, and the thickness, below which this sum is negative (the absolute instability), increases. Thus, the theory is in qualitative agreement with the paradoxical results obtained with mercury films. The most important problem is measuring the thickness of mercury films, for the purpose of quantitative checking of the theory. However, the experiments have already proved the existence of the formerly unknown electronic component of disjoining pressure, of quantum nature.

Thus, an exception has been detected to the rule which was always considered to be invariable. In accordance with this rule, the stability of symmetrical films of one-component liquids (in the absence of tensides) is impossible.

Other surprises were detected in concentrated electrolyte solutions. As appears from the theory of the electrostatic component of disjoining pressure, when the electrolyte concentration increases, the stability of symmetrical films should diminish the stronger (the higher) the charge of ions. In fact, beginning with the concentration of the order of 1 mol/l, in accordance with data from Derjaguin, Voropaeva, Rabinovich [45, 46], the equilibrium barrier preventing metal filaments from contacting one another, begins to grow instead of decreasing. In this case, the equilibrium barrier increased with the cation charge instead of decreasing (see Fig. 5). The barrier did not depend on the potential of the filaments. It should be pointed out that this does not affect the validity of the DLVO theory, since it is explicitly not applicable to such high electrolyte concentrations. The structural peculiarities of the boundary layers of water cannot be considered as the

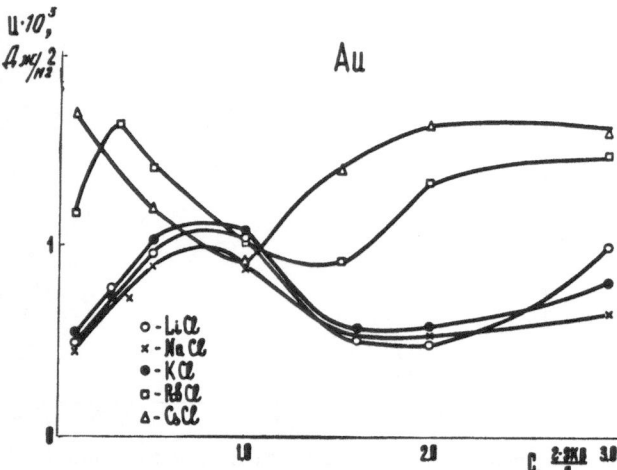

Fig. 5. The force barrier for sticking gold wires together in aqueous solutions

mechanism of the detected phenomenon, since the boundary layers are destroyed as the electrolyte concentration increases. It is apparent that a special mechanism should be the basis of this. In accordance with the graphs in Fig. 5, in this mechanism the charges of the ions exert their influence in the direction which is opposite to their role in low-concentration electrolyte solutions.

Investigation of the behaviour of blood cells offers another still more unexpected surprise. Thus, for example, Golovanov of the All Union Oncology Center [47], has shown that if blood plasma were added to a molar solution of salt (NaCl, NaBr, NaI, etc.) in a ratio of 1:5, which would give a hypertension solution, some leukocytes would become the centers of repulsion forces. Under a microscope, solution layers of about 100 µm thick, between glasses, have shown that erythrocytes are repelled by leukocytes, and that around some of these, so-called half of about 150 to 250 µm in diameter are formed, being free of erythrocytes (Fig. 6).

The formation of aureoles is complete within about 10 to 15 min. Thereafter they remain invariable for several days, as if their life activity were preserved. The diameters of aureoles depend on the kind of anions present. If erythrocytes were absent, and the number of "active" leukocytes sufficiently large, then whilst repelling one another, after a period of 1 to 2 days these would form a two-dimensional, regular hexagonal lattice with a distance between neighbouring sites of the order of 60 to 150 µm (Fig. 7). Where only part of the field is occupied by leukocytes, the periodic structure presents a sharp boundary (Fig. 8). This shows that at a certain distance the repulsion forces probably go over into the attraction ones.

It will be very important to note that not all the leukocytes by far become the active centers for the repulsion forces. In the normal blood they rarely occur. However, the portion of active centers is markedly (by several orders of magnitude) increased in the blood of persons suffering from leukosis.

At first, it was supposed that the repulsion forces stem from diffusion flows caused by leukocytes as a

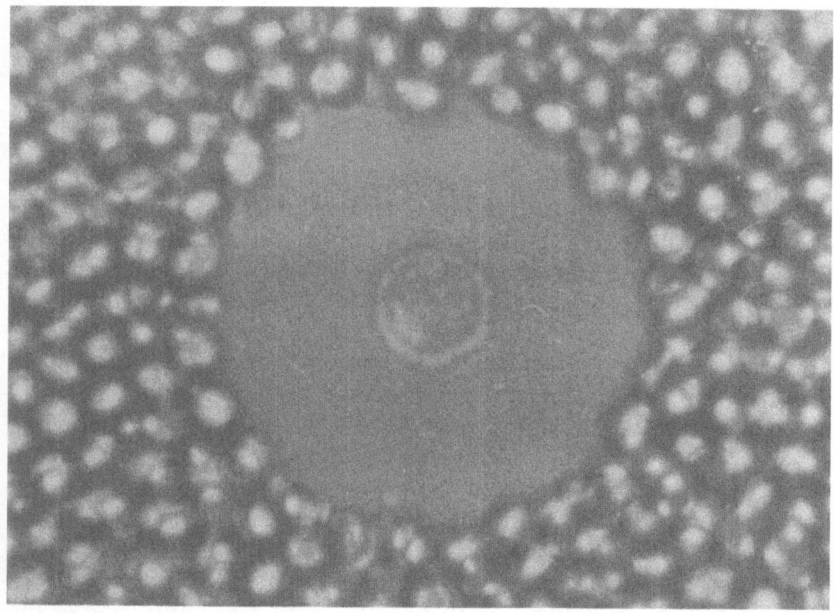

Fig. 6. Halo around leukocyte in a hypertonic solution of blood plasma

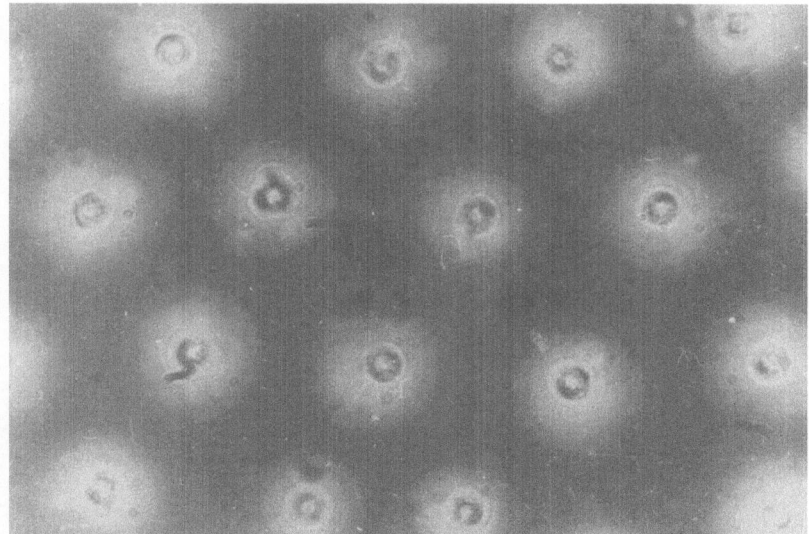

Fig. 7. The periodic structure of leukocytes

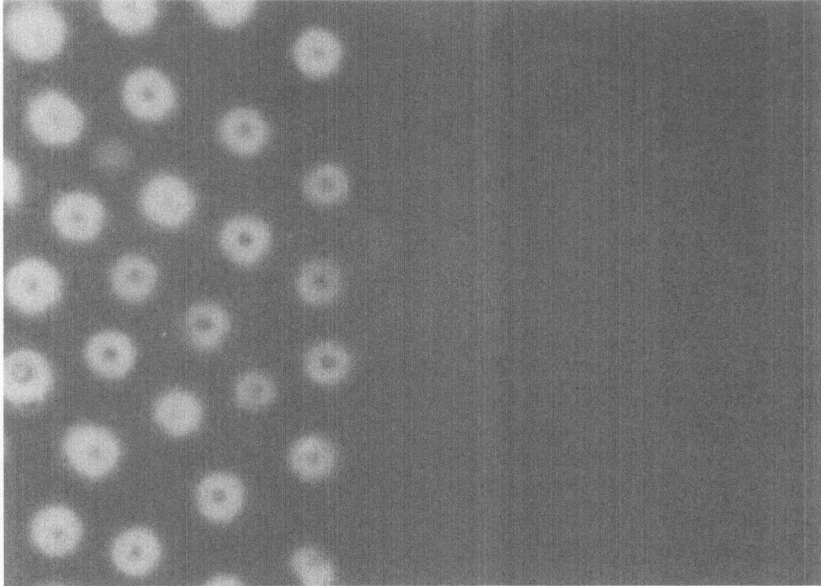

Fig. 8. The boundary of the periodic structure of leukocytes

result of the exchange processes associated with their life activity, reinforced by the flux of a hypertension salt solution.

As had been detected by Derjaguin in 1947 [48] and investigated later [49, 50], diffusion flows cause the diffusiophoresis phenomenon – the motion of suspended particles, similar to electrophoresis – which is the motion of particles under the effect of the electric potential gradient, detected 140 years earlier by Reuss. To check this mechanism, experiments were carried out in which a blood layer in the hypertension solution of blood was partitioned by a mechanical obstacle, glass filament or a platinum wire about 30 to 50 µm in diameter.

As appears from the micrograph in Fig. 9, the obstacle (barrier) does not prevent the repulsion forces from propagating at the same distance. The unique, possible clarification consists in the repulsion forces reducing to electromagnetic waves that are able to envelop mechanical barriers with a thickness of the order of 30 to 50 µm. It is obvious that the wavelength of this radiation should exceed 100 to 200 µm. According to Fröhlich [51], radiation of a biological nature is of such a wavelength, the radiation being generated by coherent excitation due to exchange processes in cells and cell membranes.

During the last 20 years, the Fröhlich idea has been developed through a long series of quantum-mechan-

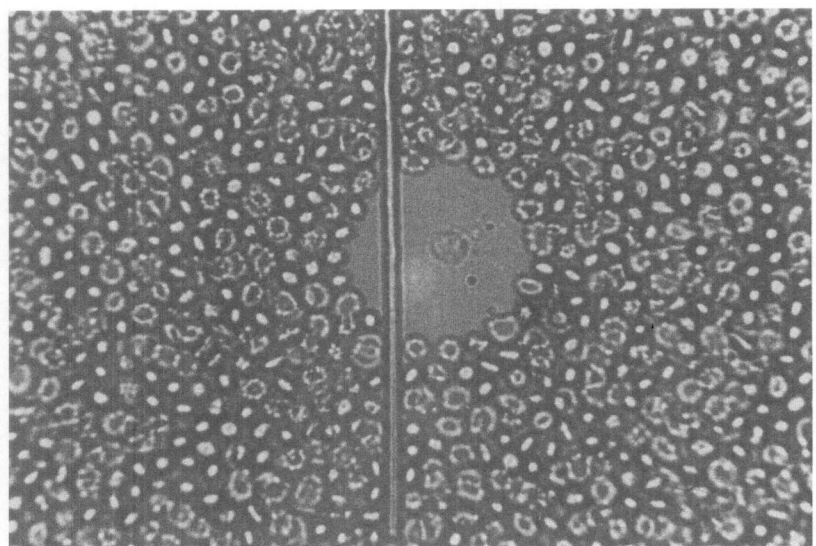

Fig. 9. The barrier in the halo around a leukocyte

ical works carried out by him and Kremer [52]. In the last few years, the idea has also been experimentally substantiated. Our work, however, constitutes a special case both with regard to the character of the phenomenon, its obviousness and regularity, and with regard to its large radius of action. Generally speaking, in dead matter the long-range effect of surface forces, which has been observed up to now, does not exceed several microns. The period of a quasi-crystalline lattice, as formed by leukocytes, is also unusually large. Until recently, the largest size of periodic colloid structures, including the biological ones, such as a tobacco mosaic, did not exceed a few microns [53].

The greatest specialist in such structures, the Japanese scientist Hachisu, has received a reprint of the paper by Golovanov and Derjaguin. Having read it, he has sent me a letter which reads: "I was surprised by the picture in which the interparticle distance was incredibly large. Surely, it was the most impressive work of 1984".

The structural component of disjoining pressure and the corresponding peculiarities of the boundary layers due to structure gained general recognition much later. These peculiarities are of immense practical significance. Let us refer to several examples.

It is generally known that in the Northern regions, even in the permafrost grounds, there exist nonfrozen water interlayers. As frequently occurs, attempts have been made to disregard the essential novelty of it, ascribing this phenomenon either to the presence of dissolved salts or to the curvature of the surface of ice formations. Another point of view, based on the struc-

tural peculiarities of nonfreezing interlayers that are in contact with the hydrophilic particles of soils, was formulated by Vershinin and Derjaguin [54]. It is noteworthy that such a "contact" melting is also observed at the ice-gas interface where, at temperatures close to 0°C, a thin water film is observed with a thickness which rapidly decreases with temperature.

The most important consequence of this observation of structural forces consists of a "thermocrystallization flow" transferring a film of unfrozen water in the direction of lowering temperature. For deriving an equation of migration of unfrozen water, Derjaguin and Churaev [55] used, on one hand, a disjoining pressure isotherm equation, and on the other, an equation of thermodynamics of steady-state irreversible transport processes, including the Onsager principle.

As a result, an equation was derived for a flow of unfrozen moisture, $q\left(\dfrac{g}{s}\right)$:

$$q = a\left(\operatorname{grad} p + \varrho_s L\,\frac{\operatorname{grad} T}{T}\right) = a\operatorname{grad} p + q_t \qquad (6)$$

where a is the coefficient of moisture conductivity of a system of pores containing the nonfrozen water, ϱ_s is the ice density, L is the ice melting heat, p is the pressure, and T is the Kelvin temperature of the medium. The second term in the right-hand side of the formula, q_t, which is proportional to the temperature gradient, expresses the thermocrystallization flow. Its considerable intensity is due to the ice melting heat,

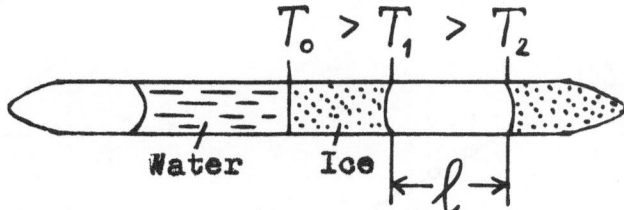

Fig. 10. Schematic diagram of an experiment involving the examination of the thermocrystallization transfer

which is included in the formula as a cofactor. For comparison, it may be indicated that the thermo-osmotic flow in fine pores is proportional to the difference in enthalpy between bulk and the boundary layers, a value which is smaller by four decimal orders.

The developed theory underwent quantitative verification in two ways. Firstly, the results of Vignes and Dijkema [56] were used. They had measured the ice growth rate, owing to the supply of water flowing from the region, where $T = 0$, through slit communications. The complete verification of the thermocrystallization transport formula is rendered difficult by the fact that the coefficient a is unknown. The method of checking realized by Churaev, Derjaguin, and others [57], is free of this disadvantage. In corresponding experiments (Fig. 10), a quartz capillary was used, half-filled with ice and half-filled with water.

A cylindrical bubble was positioned in the middle of an ice column. The mass transport velocity of ice, g/s, from a warmer meniscus to a colder one, separated from each other by the air bubble was measured. By subtracting the vapour transport velocity through the air bubble, one could obtain the thermocrystallization flow proper, with participation of only an adsorption-wetting film of water having the thickness h, coating the capillary walls around the bubble. For this film the coefficient of moisture conductivity was

$$a = \frac{2 \pi h^3 r}{3 \eta} \tag{7}$$

where η is the viscosity of the film, r is the radius of the capillary.

Therefore, in accordance with Eq. (6), we have the following expression:

$$q_t = \frac{2 \pi r h^3}{3 \eta} L \frac{\text{grad} \, T}{T} \frac{r}{c} . \tag{8}$$

The values of the film viscosity, η, were assumed to be equal to the viscosity of nonfreezing interlayers between ice and quartz, measured earlier by Barer,

Churaev and Derjaguin [58]. Hence, having experimentally determined the value of q_t, one could find a single unknown value, h, the thickness of an adsorption-wetting film in equilibrium with almost saturated water vapour at a temperature neare to $0\,°C$. It was found to be close to $100 \, \text{Å}$, which is near the values obtained earlier by Derjaguin and Zorin [11] from direct ellipsometric measurements.

In previous works by other researchers [59], suppositions of a very particular nature on the mechanism of the effect were put forward, while the essential point remained unclarified: namely, its connection with the ice melting heat.

The developed theory of the thermocrystallization transport has been simultaneously transformed into the first, strictly quantitative theory of the frost-heaving of soils – a global phenomenon creating immense difficulties in road construction and building engineering in the Northern region [60]. Underlying the frost-heaving phenomenon, setting up forces of the order of several hundred atmospheres, is a combination of non-equilibrium processes of thermal migration with the development of equilibrium disjoining pressure, basically of a structural nature. Hence, it becomes apparent how unjustified specialists in the field of colloid and surface phenomena were in not having acknowledged structural effects and forces, for a long time and in underestimating them.

Another process for which ignoring the special structure of boundary layers of liquids is also inadmissible, is the that of inverse osmosis, consisting of using the filtration of solutions through fine-porous membranes for separation of their components, as is done, for example, in the desalting of water. As has been shown by Churaev and Derjaguin [61, 62], the most important factor in the selective action of "inverse osmosis" is just a reduced dissolving power of the boundary layer when aqueous solutions pass through hydrophilic membranes.

Here it would be appropriate to note that since the membranes employed are penetrable to all the components, it would be more correct to describe their use as "inverse capillary osmosis" rather than inverse osmosis. As has been demonstrated in the work by Derjaguin and Koptelova (Milekhina) [63], the concentration gradient of a dissolved component (including a non-electrolyte) across a porous membrane causes a solution flow through it, which is proportional to the moment of the component adsorption on the surfaces of pores, y, the absorption moment being equal to:

$$y = \int_0^\infty \Gamma(h) h \cdot dh \tag{9}$$

where $\Gamma(h)$ is the density of the adsorption atmosphere at a distance h from the wall surface.

That such a capillary osmosis occurs not only in electrolyte solutions, but also in those of nonelectrolytes, proves that in the latter case, too, the adsorption layer has a diffuse structure. Still earlier, in 1947, Derjaguin [48] detected a cognate phenomenon of diffusiophoresis — motion of particles suspended in a solution under the effect of a concentration gradient. The diffusiophoresis rate is also proportional to the adsorption moment. This is based on the fact that the capillary osmosis and diffusiophoresis velocities are due to the slipping of the solution over the wall, which is proportional to the adsorption moment [64]. Between diffusiophoresis and electrophoresis, which was discovered by Reuss in 1807, there exists a similarity, as well as between capillary osmosis and electroosmosis.

In distinction from the electrokinetic phenomena, however, in accordance with Derjaguin, Dukhin, Ul'berg and Dvornichenko [65], capillary osmosis and diffusiophoresis are observed to occur both in the absence of ions and at an isoelectrical point of ionic solutions. Therefore, both phenomena may serve as methods for investigation of nonionic diffuse adsorption atmospheres.

The developed theory also proved suitable for explaining, in terms of positive adsorption, the phenomenon of an anomalous capillary osmosis making the solution flow in the direction of a higher concentration. The major importance of inverse osmosis in membranes (in addition to capillary osmosis) shows the genuine significance of the new concept on the special structure of boundary layers of liquids.

The fact that the effect, depending on the discrete structure of liquid, and leading to oscillations with a period equal to the diameter of molecules, and the long-range effects, are referred to under the common term of structural effects also interferes with achieving complete clarity. As calculations carried out by a numerical experiment and molecular dynamics method [66] suggest liquids with approximately spherical molecules form a multilayer structure with a thickness of the order of molecule diameters near to an absolutely rigid and smooth wall. Such a structure is similar to that of soap films, as detected by Perrin and Wals at the beginning of the twentieth century.

At that time, that was the way in which to assess the size of molecules. At the present time, this is already devoid of essential novelty. Moreover, such a structure is not realized at soft interfaces, as for example, at an interface with gas (with the exception of tensides). It would be expedient to designate these structural

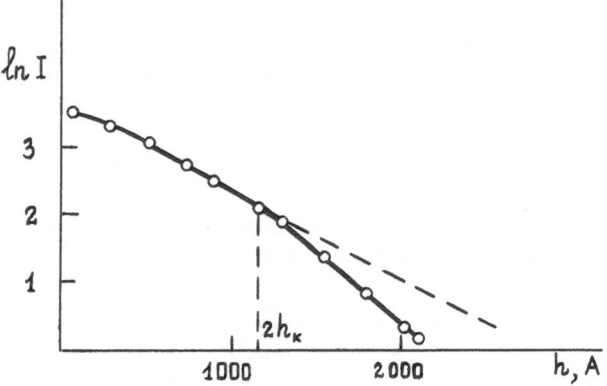

Fig. 11. Dependence of the light decay on the thickness of a nitrobenzene interlayer

peculiarities and effects as microstructural ones. As distinct from effects dependent on the structure resulting from direct or relay long-range action of an interface, weakening at distances of the order of many molecular diameters (without oscillations, depending on the close-range repulsion of electron shells, which determines the molecule size).

As early as in 1952 [67], the author put forward a supposition that in certain cases a special structure of boundary layers, different from the bulk one, might be approximately uniform, transforming jumpwise into the bulk structure at a certain distance from the substrate. That such a concept is adequate was proved in the most direct, trustworthy and conspicious manner in 1980—82 by Popovsky, Altoiz and Derjaguin [68]. These experiments involved investigation of the absorption of polarized light in a wedge-shaped layer of a number of organic liquids, which was formed between two quartz platelets (Fig. 11). The measurements have also revealed that in layers arranged at a distance of less than $h \approx 500$ Å from the substrate, a uniform, liquid-crystalline structure is observed which is characterized by birefringence, Δn, and a parameter of the order S. The corresponding data for the liquids tested are presented in Table 1. Even earlier, Popovsky [69] and Derjaguin had measured the heat of transition of a boundary phase into the bulk one, λ, as a function of temperature. This heat of transition is close to the heat of the transition from ordinary liquid-crystalline phases into the bulk one.

The theory [70] enabled one to explain completely all these observations, reducing them to the effect of three factors; namely, the contact close-range action of the substrate, the molecular interactions inside liquid layers, and the surface long-range forces guaranteeing the stability of a layer of thickness h.

As is known from a number of works, in the case of water and several other polar liquids, the boundary layers exhibit a nonuniform structure, and the structural forces vary with thickness according to an exponential law without any jump whatsoever [71].

However, the phenomenon of thermoosmosis exists, demonstrating a complex distribution of the excess enthalpy density, depending on structural peculiarities in thin interlayers of water and other liquids. This phenomenon essentially results in a steady-state flow of liquid through a porous partition wall under the effect of a temperature gradient between both sides of the partition wall. We had investigated this in as early as 1941 [72].

Application of the Onsager principle to the porous partition wall gives a relationship:

$$Q_P = \frac{W_T}{\mathrm{grad}\, P} \frac{\mathrm{grad}\, T}{T} \tag{10}$$

where Q_P is the thermoosmotic flow under isobaric conditions and under the influence of a temperature gradient at a Kelvin temperature T; W_T is the isothermal transfer heat under the effect of a pressure gradient generating a filtration flow:

$$Q_0 = -\beta\, \mathrm{grad}\, P .$$

It is obvious that W_T is equal to:

$$W_T = \iint \Delta H v\, ds \tag{11}$$

where ΔH is the excess enthalpy density in the pores of the partition wall, where the filtration flow velocity, v, and integration extend to all the areas of the normal cross-section of pores.

Now the filtration flow is equal to

$$W_0 = \iint v\, ds = -\beta\, \mathrm{grad}\, P \tag{12}$$

where β is the filtration coefficient; h is the thickness of the porous partition wall. From Eqs. (1) and (3) it follows:

$$Q_P = -\frac{\iint \Delta H v\, ds}{\iint v\, ds} \beta \frac{\mathrm{grad}\, T}{T} . \tag{13}$$

Vozny and Churaev [73] measured the thermoosmotic flows through glasses of the same composition and porosity (and hence, of the same total cross-section of pores), but different in the pore radii. In Fig. 12 corresponding graphs are presented.

In a sample (No. 5) possessing wide pores, thermoosmosis is directed to the hot side which, in accordance with Eq. (11), indicates that a reduced specific enthalpy is predominant in the boundary layers that do not overlap in wide pores.

In the pores up to 550 Å radius, the direction of thermoosmosis is reversed. Consequently, in these pores the mean, effective value of ΔH_e

$$\Delta H_e = \frac{\iint \Delta H v\, ds}{\iint v\, ds}$$

becomes positive.

The thermoosmosis rate increases as the pore radius decreases further. This means that the value of ΔH_e

Table 1.

Preparation	λ_i nm	μ_{iso} mkm^{-1}	μ_s mkm^{-1}	S	d_s nm
Nitrobenzene *) $C_6H_5O_2$	262	20.7	14.7	0.27	55
Aniline $C_6H_5H_2$	285	3.24	2.3	0.29	59
Acetophenone, $C_6H_5COCH_3$	278	2.3	1.7	0.31	62
Parachloro-toluene, $CH_3C_6H_5Cl$	220	21.6	15.5	0.28	53
Ethyl ester of benzoic acid, $C_6H_5\text{-}COOC_2H_5$	225	19.2	16.5	0.14	47
Octyl ester of benzoic acid, $C_6H_5COOC_8H_{17}$	225	14.8	11.2	0.24	62.5

*) An additional lyophilization of the quartz surface under the effect of hydrogen flame resulted in an increase both in the equilibrium thickness of the boundary phase up to $2d_s = 130$ nm and in the value of S up to 0.32.

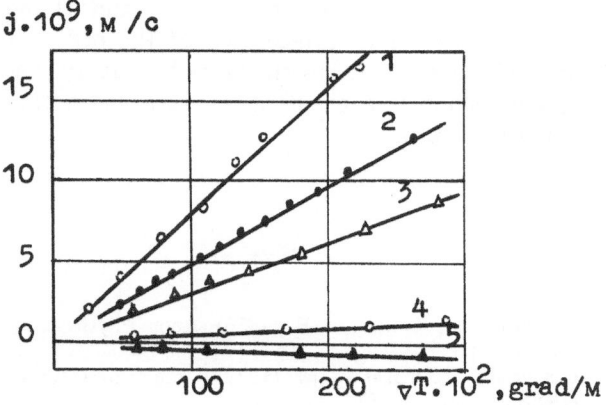

Fig. 12. Dependence of the thermoosmotic flow rate on the capillary radius. r: $1 - r = 45$ Å, $2 - r = 80$ Å, $3 - r = 100$ Å, $4 - r = 550$ Å, $5 - r = 15,000$ Å

increases abruptly, compensating for a decrease in the filtration coefficient (which is proportional to the mean square of the pore radius).

These results may be attributed to the fact that the peripheral portions of the boundary layers possess a reduced enthalpy density, whereas in the near-to-wall parts there is an enhanced enthalpy density, which increases markedly as the pores walls are approached. It is surprising, however, that the sign of ΔH_e is reversed in the region of pores with radii of the order of 10^3 Å. It is possible that, owing to a special sensitivity of thermoosmosis to structural peculiarities, it is possible to detect these where other methods are unsuitable for this.

That the values of ΔH_e, found by observing thermoosmosis, are characteristic of the special structure of boundary layers, follows from the fact that for water ΔH_e disappears at a temperature of 70 °C, when all other peculiarities of aqueous boundary layers due to structure disappear, too. Such a marked increase in the value of ΔH_e in the range radii of the order of 100 Å is also surprising.

Perhaps this is a result of the overlapping of boundary layers, which enhances their structural peculiarities in a nonadditive manner. All the aforesaid indicates that thermoosmosis is an extremely sensitive and specific means of examination of the special structure of boundary layers.

In conclusion, it is important to state that the fundamentals of the science of surface forces have reached an extensive use in developing a quantitative theory of the mass and heat transfer in soils and ground, which is not limited merely to general notions of hydromechanics and heat conductivity. In this connection, the fruitful research by Nerpin [74] should be referred to in particular. An interesting example of this is the accumulation of immense resources of water at a depth of around 200 to 300 m under sand of the Karakum desert, of a surface area of the order of 10^4 km². Chubarov [75] has shown in his theory that this phenomenon is a result of the flow of wetting films fed by rare rains.

References

1. Derjaguin B, Obukhov E (1936) Acta Physicochim URSS 5, 1:1
2. Derjaguin B, Kusakov M (1939) Acta Physicochim URSS 10, 1:25; 2:153
3. Derjaguin B (1955) Kolloidn Zh 17, 3:207; (1975) Colloid Polym Sci 253, 6:492; (1978) J Colloid Interface Sci 66, 3:389
4. Derjaguin B, Titievskaya A (1953) Kolloidn Zh 6:416; (1960) 22, 4:398; (1954) Discuss Faraday Soc 18:27; (1957) Proc II Intern Congr Surface Activ, London, Vol 1, p 211; Sheludko A (1967) Adv Colloid Interface Sci 1, 4:391
5. Derjaguin B, Abriskosova I, Leib F (1951) Vest Akad Nauk SSSR 6:125; Derjaguin B, Abrikosova I (1951) Zh Eksp Teor Fiz 21, 8:945; (1956) ibid. 30, 6:993; (1956) ibid. 31, 1:3; (1953) Dokl Akad Nauk SSSR 90, 6:1055
6. Lifshitz E (1954) Dokl Akad Nauk SSSR 97, 4:643; ibid. 100, 5:879; (1955) Zh Eksp Teor Fiz 29, 1:94; Derjaguin B, Abrikosova I, Lifshitz E (1958) Usp Fiz Nauk 64, 3:493
7. Derjaguin B (1938) Izv Akad Nauk SSSR, Ser Khim 5:1153; (1939) Acta Physicochim URSS 10, 3:333 (1940) Kolloidn Zh 6, 4:291; (1941) ibid. 7, 3:285; (1979) Uspekhi Khimii 48, 4:675
8. Derjaguin B, Landau L (1941) Acta Physicochim URSS 14, 6:633; (1941) Zh Eksp Teor Fiz 11, 12:802; (1945) ibid. 15, 11:663 Parfitt GG, Tideswell M (1981) J Colloid Interface Sci 79, 2:518
9. Derjaguin B (1954) Discuss Faraday Soc 18:85; Usui S (1972) In: Progress in surface and membrane science. Academic Press, New York, Vol 5, p 223; (1973) J Colloid Interface Sci 44, 1:107 Devereux C, de Bruyn P (1963) Interaction of plane-parallel double layers. MIT Press Cambridge (Mass.), p 361; (1962) J Chem Phys 37, 9:2147; (1964) J Colloid Sci 19, 3:302; Pugh R, Kitchener J (1971) J Colloid Interface Sci 35, 3:656; Wiese G, Healy T (1970) Trans Faraday Soc 66, 3:490; Smilga V (1960) Kolloidn Zh 22:615; Bleier A, Matijevic E (1976) J Colloid Interface Sci 55, 3:5 – 10; (1978) J Chem Soc Farad Trans, Part I, 74, 9:1346
10. Derjaguin B, Dukhin S (1961) In: Bull Inst Min Met, Transactions, 1960 – 1961, part 5, 651, 70:221; (1984) Kinetic Theory of Flotation of Small Particles. In: Surface Colloid Sci 13, edit. Matijevic E, Plenum Press, New York London, p 17
11. Derjaguin B, Zorin Z (1955) Zh Fiz Khim 29, 10:1755. (1957) In: Proc II Intern Congr Surface Activ, London, Vol 2, p 145; (1977) Hu P, Adamsson A (1977) J Colloid Interface Sci 59, 3:605; Pashley R, Kitchener J (1974) J Colloid Interface Sci 49, 2:249
12. Derjaguin B, Churaev N (1974) J Colloid Interface Sci 49, 2:249
13. Derjaguin B (1940) Zh Fiz Khim 14, 2:137
14. Bangham D, Mosallam S, Saveris Z (1938) Trans Farad Soc 34:554; (1946) J Chem Phys 14, 5:352

15. Karasev V, Derjaguin B (1978) Dokl Akad Nauk SSSR 62, 6:761;
(1955) ibid. 101, 2:289;
Derjaguin B, Karasev V et al (1969) Dokl Akad Nauk SSSR 187, 4:846;
(1971) In: Thin liquid films and boundary layers, Academic Press, New York London, p 98;
(1975) Research in surface forces. Consult Bur, New York, Vol 4, p 188;
(1958) Wear, Vol 1, 4:277
16. Derjaguin B, Zakhavaeva N et al (1957) In: Proc II Intern Congr Surf Activ, London, Vol 2, p 531;
(1966) In: Research in surface forces , Cons Bur, New York, Vol 2, p 156
17. Churaev N et al (1971) In: Thin liquid films and boundary layers, Academic Press, New York London, p 213;
Derjaguin B et al (1971) In: Research in surface forces, Cons Bur, New York, Vol 3, p 261;
(1974) In: Surface Forces in Thin Films and Stability of Colloids, Moscow, published in Russian by Nauka, p 90
18. Green-Kelly, Derjaguin B (1964) Trans Faraday Soc 60:449;
(1966) In: Research in surface forces, Cons Bur, New York, Vol 2, p 115
19. Derjaguin B, Sidorenkov G (1941) Dokl Akad Nauk SSSR 32, 32:622;
Derjaguin B, Melnikova M (1958) Highway Res Board, Spec Rep 40:43;
Derjaguin B, Ershov A, Churaev N (1975) In: Research in surface forces, Cons Bur, New York London, Vol 4, p 171
20. Derjaguin B, Karasev V, Chromova E (1971) In: Research in surface forces, Cons Bur, New York, Vol 3, p 25;
(1980) J Colloid Interface Sci 78, 2:274;
(1986) 109, 2:586
Derjaguin B (1983) In: Surface Forces and Boundary Layers of Liquids, Nauka, Moscow, p 3;
Derjaguin B, Karasev V, Khromova E (1980) Kolloidn Zh 42, 4:808;
Derjaguin B et al (1987) Colloids and surfaces, in press
21. Allen L, Matijevic E (1969) J Colloid Interface Sci 33, 3:420;
Klimentova Yu, Kirichenko L, Vysotsky Z (1975) In: Research in surface forces, Cons Bur, New York London, Vol 4, p 77;
Chernoberezhsky Yu et al (1979) In: Surface Forces in Thin Films, published in Russian by Nauka, Moscow, p 67;
Frolov Yu et al (1983) Kolloidn Zh 45, 3:509
22. Low P (1980) Soil Sci Amer J 44, 4:667;
(1982) J Colloid Interface Sci 89, 2:366;
(1983) J Colloid Interface Sci 96, 1:229
23. Rabinovich Ya, Derjaguin B, Churaev N (1982) Adv Colloid Interface Sci 16:63
24. Peschel G, Belouschek P (1976) Prog Colloid Polym Sci 60:108–119;
(1977) Z Phys Chem NF 108, 2:145;
(1982) Colloid Polymer Sci 260, 4:44
25. Israelachvili J (1978) Farad Discuss Chem Soc 65:20;
(1978) J Chem Soc Farad Trans Part I, 74, 4:975;
See also: Derjaguin B, Churaev N (1974) J Colloid Interface Sci 49, 2:249
    a. Israelachvili J (1981) Philos Mag A 43, 3:753
26. Yaminsky VV, Yushchenko VS, Amelina EA, Shchukin ED (1983) J Colloid Interface Sci 96, 2:301–306; 307–314
27. Dzyuloshinsky I, Lifshitz E, Pitaevsky A (1959) Zh Eksp Teor Fiz 37, 1:229
28. Derjaguin B et al (1965) Discuss Farad Soc 40:246
29. Derjaguin B et al (1947) Kolloidn Zh 9, 5:335
30. Derjaguin B, Koptelova M (1975) In: Research in surface forces, Cons Bur, New York London, Vol 4, p 182;
(1977) Kolloidn Zh 39, 6:1060
31. Derjaguin B (1980) Colloid Polym Sci 258, 4:433;
Kusakov M, Titievskaya A (1940) Dokl Akad Nauk SSSR 28, 4:333
32. Sheludko A, Ekserova D (1961) Godishn Sofiisk un-t, Khim fak 54, 3:205;
Derjaguin B, Voropaeva T, Kabanov B (1964) J Colloid Interface Sci 19, 2:113
33. Derjaguin B, Churaev N (1982) J Colloid Interface Sci 87, 2:543
34. Kuni F, Rusanov A, Brodskaya E (1975) In: Research in surface forces, Cons Bur, Vol 4, p 240
35. Vasil'ev Kh, Ivanov I (1979) Godishn Sofiisk un-t, Khim fak 60, 1:111
36. Gorelkin V, Smilga V (1972) Kolloidn Zh 34, 5:685;
(1973) Dokl Akad Nauk SSSR 208:635
37. Davies B, Ninham B (1972) J Chem Phys 56, 12:5797
38. Mitchell D, Richmond P (1974) J Colloid Interface Sci 46, 2:128
39. Yaroshchuk A, Dukhin S (1983) Kolloidn Zh 45, 3:527;
Dukhin S, Shilov V (1983) Usp Koll Khim, Kiev, Naukova Dumka, p 96
40. Krasnokutskaya M, Glasman Yu (1971) In: Research in surface forces, Cons Bur, New York, Vol 3, p 189;
Blashchuk Zh, Glasman Yu (1975) ibid. Vol 4, p 72;
Glazman Yu (1979) Croat Chem Acta 52, 2:115;
(1982) J Dispers Sci Technol 3, 1:67
41. Derjaguin B, Dukhin S, Yaroshchuk A B (1984) Kolloidn Zh 46, 2:225;
(1987) J Colloid Interface Sci 115, 1:234
42. Derjaguin B, Leonov L, Yashin V (1983) Dokl Akad Nauk SSSR 273, 1:122;
Derjaguin B, Leonov L, Roldughin V (1985) Colloid Interface Sci 108, 1:207;
Derjaguin B, Roldughin V (1985) J Surface Sci 159:69
43. Derjaguin B, Roldughin V (1983) Dokl Akad Nauk SSSR 270, 3:642;
(1985) Kolloidn Zh 47, 2:322
44. Derjaguin B, Martynov G (1962) Dokl Akad Nauk SSSR 144, 4:825
45. Voropaeva T, Derjaguin B, Kabanov E (1963) Research in surface forces, Cons Bur, New York, Vol 1, p 116;
Derjaguin B, Voropaeva T (1964) J Colloid Sci 19:113
46. Derjaguin B, Rabinovich Ya (1969) Kolloidn Zh 31, 1:47
47. Derjaguin B, Golovanov M (1984) J Colloid Surfaces 10, 1:77;
(1979) Kolloidn Zh 41, 4:649;
(1983) Dokl Akad Nauk SSSR 272, 2:479;
(1986) Kolloidn Zh 48, 2:248
48. Derjaguin B et al (1947) Kolloidn Zh 9, 5:335
49. Derjaguin B, Dukhin S (1964) Dokl Akad Nauk SSSR 159, 2:401; 3:636
50. Derjaguin B et al (1983) In: The Collection "Surface Forces and the Boundary Layers of Liquids", Moscow, published in Russian by Nauka, p 84;
(1986) Kolloidn Zh 48:4

51. Fröhlich H (1968) J Quant Chem 2:641;
(1970) Nature 228:1093;
(1975) Physics Letters 51A:21
52. Fröhlich H, Kremer F (eds) (1983) Coherent Excitat in Biolog Syst. Proc in Life Sci, Springer, Berlin Heidelberg
53. Efremov I (1976) Periodic Colloid Structures. In: Surface Colloid Sci, Wiley, New York, Vol 8, p 85
54. Vershinin P, Derjaguin B (1949) Izv Akad Nauk SSSR, Ser Geogr Geofiz 13, 2:108
55. Derjaguin B, Churaev N (1978) J Colloid Interface Sci 67, 3:391;
(1986) Cold Regions Sci Technol 12, p 57;
(1980) Kolloid J (Russ) 42, 5:842
56. Vignes M, Dijkema K (1974) J Colloid Interface Sci 49, 2:165
57. Derjaguin B, Churaev N et al (1981) J Colloid Interface Sci 84, 1:182
58. Barer SS, Churaev NV, Derjaguin BV et al (1980) J Colloid Interface Sci 74, 1:173
59. Vignes-Adler M (1977) Colloid Interface Sci 60, 1:162
60. Dostavalov B, Kudryavtsev V (1967) Obshchee merzlotovedenie, published by MGU
61. Derjaguin B et al (1985) Dokl Akad Nauk 285, 1:140
62. Churaev NV, Derjaguin BV (1986) Khim Tekhnol 2:180
63. Derjaguin B, Koptelova M (1969) Kolloidn Zh 31, 5:692;
(1975) Research in surface forces, Cons Bur, New York London, Vol 4, p 182
64. Derjaguin B, Dukhin S (1976) Elektroforez, Moscow, Nauka
65. Derjaguin B, Dukhin S Z Ul'berg, Dvornichenko G (1987) Kolloidn Zh 48:4
66. Antonchenko (1983) The Microscopical Theory of Water in the Pores of Membranes, published in Russian by Naukova Dumka, Kiev
67. Derjaguin B (1952) The proceedings of the Union Conference on the Colloid Chemistry, Kiev (in Russian) Akad Nauk SSSR, p 26
68. Popovsky Yu, Altoiz B (1981) Kolloidn Zh 43, 6:1177;
Derjaguin B, Popovsky Yu, Altoiz B (1982) Dokl Akad Nauk SSSR 262, 4:853;
Derjaguin BV, Popovsky Yu M (1982) Kolloidn Zh 44, 5:863
69. Popovsky Yu, Derjaguin B (1964) Dokl Akad Nauk SSSR 159, 4:897;
(1967) ibid 175, 2:385
70. Derjaguin BV, Popovsky Yu M, Altoiz BA (1983) J Colloid Interface Sci 96, 2:492;
Derjaguin BV (1986) Fluid Interfac. Phenom. Croxton CA (ed) Wiley, p 739
71. Churaev NV, Derjaguin BV (1985) J Colloid Interface Sci 103, 2:542;
(1986) Fluid Interfac. Phenom. Croxton CA (ed) Wiley, p 663
72. Derjaguin B, Sidorenkov G (1941) Dokl Akad Nauk SSSR 2, 9:622
73. Vozny P, Churaev N (1977) Kolloidn Zh 39, 2:264; 3:438
74. Nerpin S, Chudnovsky (1967) Physics of Soil, Moscow, published in Russian by Nauka
75. Chubarov V (1972) Pitanie gruntovykh vod peschanoi pustyni cherez zonu aeratsii, Moscow, Nedra

Author's address:

B. V. Derjaguin
Institute of Physical Chemistry
USSR Academy of Sciences
Leninskij Prospect 31
117915 Moscow, USSR

Progress in Colloid & Polymer Science   Progr Colloid & Polymer Sci 74:31–37 (1987)

The swelling of lamellar phases in oil: The role of double layer interactions

H. Wennerström

Division of Physical Chemistry 1, Chemical Center, Lund, Sweden

Abstract: The electrostatic contribution to the force across the apolar region in a lamellar liquid crystal is considered. For an ionic surfactant consisting of one small hydrophilic ion and one larger hydrophobic ion, a microscopic charge separation occurs. If the hydrophobic ion has a finite solubility in the apolar region a double layer force will be generated across the apolar region. It is shown that under realistic conditions this force can overcome the attractive van der Waals force and it is suggested that the extensive swelling of a lamellar phase in hydrocarbon, recently reported by Larché et al. [11] might be explained in terms of this double layer repulsion. It is also concluded that this type of double layer forces can be of important for the stability of microemulsions and water in oil emulsions.

Key words: Double layer interaction, swelling, lamellar phase

Introduction

Electrical double layer interactions can induce an extensive swelling in water of lyotropic lamellar liquid crystals [1–4]. The effect is particularly dramatic if the aqueous solubility of the ionic amphiphile is small and if the water is free from electrolytes [5, 6]. In such a case the double layer force decays slowly with separation between the charged surfaces and only a small surface charge density is needed to overcome the attractive van der Waals force [7]. It has been demonstrated that it is possible to account quantitatively for this swelling using the non-linearized Poisson-Boltzmann equation to describe the electrostatic effects [8–10].

Recently it was observed by Larché et al. [11] that a lamellar phase in the system water-sodium octylbenzene sulfonate (SOBS)-pentanol could incorporate large quantities of alkane retaining the lamellar structure. Similar less striking observaqtions of incorporation of alkanes have previously been made by several groups [12–14]. The very extensive swelling found by Larché et al. up to a repeat distance of 250 nm, shows that a long range repulsive force is operating in the system. To explain their own experimental results, di Meglio et al. [12] have invoked a mechanism with a short range molecular repulsive interaction which leads to a quenching of large scale undulations, a mechanism originally discussed by Helfrich [15] for the lecithin system. In this communication we analyze an alternative mechanism that might account for these experimental observations.

The lamellar liquid crystal

In a liquid crystal formed by a system water-ionic surfactant-alcohol-alkane one can identify three molecularly different regions [10, 16] (see Fig. 1); (1) The aqueous region where, apart from the water, the counterions reside, Na^+ in the case of SOBS. The aqueous solubility of the surfactant ion can be ignored except at the highest water contents [10]. (2) The interfacial region containing surfactant ions and alcohol as a cosurfactant. Possibly one could also include some water and 'bound' counterions in this region. (3) The apolar region containing the alkane and the alcohol. This region also contains small amounts of water and, as will be argued below, surfactant ions in small amounts.

Figure 1 shows an average ideal structure, while in reality there will be capillary waves at the interface making it locally nonplanar. In contrast of Helfrich's theory, in which these waves play a fundamental role, we will ignore them for simplicity and only note that their effect will be to modify the distance dependence of the calculated force [17, 18].

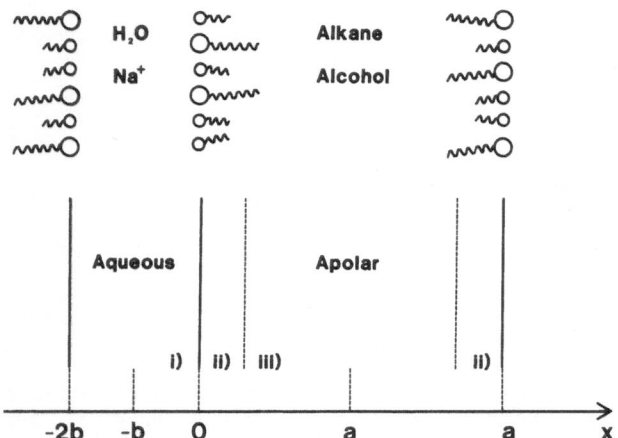

Fig. 1. Schematic model of the lamellar liquid crystal. (i) the aqueous region, thickness 2b, (ii) the interfacial region and (iii) the alkane. 2a is the thickness of the combined alkane and interfacial regions

The experimental finding of Ref. [11] is that when an alkane-alcohol mixture is added to the lamellar phase, it simply swells under suitable conditions. This implies that the chemical potential of the alkane and the alcohol is lower in the lamellae than in the isotropic liquid added. In effect, by definition, there exists a repulsive interaction across the apolar region. In such a system there is always an attractive van der Waals force, the magnitude of which could be estimated by the formula [19]

$$F/A = \frac{H}{48\pi} \left(\frac{1}{(a)^3} - \frac{2}{(a+b)^3} + \frac{1}{(a+2b)^3} \right) \qquad (1)$$

where H is the Hamaker constant. The repulsive interaction must then be stronger than this attractive force. The extreme long range character of the force suggests that it is of electrostatic origin. In the following sections this suggestion is analyzed in detail.

The electrostatic interactions in a lamellar liquid crystal

As discussed in the previous section, the small hydrophilic counterions reside exclusively in the aqueous region, being excluded from the apolar region due to the unfavourable solvation conditions there. Conversely, the amphiphilic ion is practically excluded from the aqueous region, due to the hydrophobic interaction. There is thus an inherent mechanism in the system leading to local separation of charges. The simplest way to model the electrostatic effect is, then,

to take the local 'solvation' interactions into account by constraining the different ionic species to the aqueous region for the hydrophilic ions and to the other two regions for the amphiphilic ions. For the present we will make no distinction between regions (ii) and (iii) and will postpone the discussion of the difference between these two regions until the following section of the paper.

In solving for the ion distribution in the system the Poisson-Boltzmann (PB) equation is used. With the planar geometry and with only one type of species in each region, the PB equation has a simple solution [20, 21]. In the aqueous region, the electrostatic potential Φ is

$$\Phi(x) = \frac{2kT}{z_c e} \ln \{\cos [s_w(1+x/b)]\} \qquad (2)$$

where $z_c e$ is the charge of the counterions. The potential has, by convention, been chosen as zero in the mid-plane of the aqueous lamella ($x = -b$). The detailed variation of the potential is determined by the boundary conditions through the value of the dimensionless parameter s_w given by

$$s_w \tan(s_w) = \frac{(z_c e)^2 (n/A) b}{2kT\varepsilon_w \varepsilon_0} \equiv K_w \qquad (3)$$

where n/A is the number of ions per unit area and ε_w the relative dielectric permittivity of the aqueous region. The ion distribution $c_w(x)$ is, $-b \geq x > 0$

$$c_w = 2kT\varepsilon_w \varepsilon_0 \frac{s_w^2}{(z_c e b)^2} \frac{1}{\cos^2\{s_w(1+x/b)\}} . \qquad (4)$$

For the distribution of amphiphilic ions, charge $z_a e$, in the regions (ii) and (iii) an analogous solutions applies, and

$$\Phi(x) = (2kT/z_a e) \ln \{\cos [s_a(1-x/a)]\} + \Phi' \qquad (5)$$

$$\Phi' = (2kT/z_a e) \ln \{\cos (s_a) \cos (s_w)\} \qquad (6)$$

assuming $z_c = +1$ and $z_a = -1$.

The parameter s_a is determined as in Eq. (3), except that the dielectric permittivity is changed to the value ε_a in the apolar region. The ion concentration profile is for $0 < x \leq a$, in analogy with Eq. (4),

$$c_a(x) = 2kT\varepsilon_a \varepsilon_0 (s_a/z_a e b)^2 / \cos^2 [s_a(1-x/a)]. \qquad (7)$$

In Fig. 2 the concentration profiles are shown for a typical set of parameters. A notable feature of this figure is that the strong electrostatic counterion 'bind-

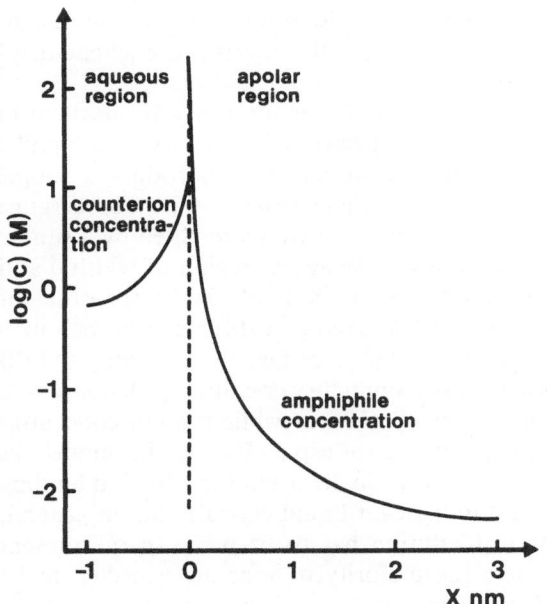

Fig. 2. Calculated ion concentration profiles. Counterions are excluded from the apolar region and amphiphile ions from the aqueous region. Note the logarithmic concentration scale. Parameters used: $\varepsilon_w = 80$, $\varepsilon_a = 4$, $z_w = +1$, $z_a = 1$, $b = 10$ Å, $a = 30$ Å, $\sigma = 0.2$ cm^{-2}

ing', well established for the hydrophilic ions [22, 23], is even more pronounced for the amphilic ions due to the low value of ε_a. In fact, more than 99% of the amphiphilic ions are located in the interfacial region indicated in Fig. 1. It is commonly assumed that in a system as in Fig. 1 the amphiphilic ions are located exclusively at the interface, due to a specific attraction. What the electrostatic calculation shows is that the strong accumulation of the ions in the interfacial region can be accounted for simply by the long-range electrostatic forces, but in this case there is a small amount of ions dissolved into the medium.

Without the specific interaction in the interfacial region, there will be a slight overlap between the double layers associated with the two interfacial regions of one apolar lamella. The repulsive force per unit area is [8, 24]

$$F/A = kTc(a) \tag{8}$$

and the ion concentration in the center is from Eq. (7)

$$c(a) = 2kT\varepsilon_a\varepsilon_0(s_a/z_aea)^2. \tag{9}$$

With no specific binding to the surface the average ion density n/A per unit area is so high that the constant

K of Eq. (3) is much larger than unity, making $s_a \approx \pi/2$. The electrostatic pressure acting across the apolar lamella is then

$$F/A = \varepsilon_a\varepsilon_0/2 \cdot (\pi kT/ea)^2. \tag{10}$$

The force is clearly long-range-decaying more slowly than the van der Waals' force and is stronger than this force at distances of practical relevance. The conclusion then, is that a lamellar system of this type with substantial amounts of apolar compounds is stable, relative to a system with the apolar compounds in a separate isotropic phase.

This conclusion is, at present, based on the assumption that the amphiphilic ion does not have a special preference for the interfacial region, which in most cases is an oversimplification of the problem.

The electrostatic force in the presence of specific interactions at the interface

Amphiphilic ions are located at the interface through a complex interplay between different effects. The hydrophobic interaction anchors the amphiphile in the apolar region, while the ionic group clearly prefers the aqueous medium. Above, we have modelled the media through the dielectric permittivity with a sharp discontinuity at $x = 0$ (see Fig. 1). One can think of more realistic models but they become increasingly difficult to handle. Given that the detailed molecular understanding of the interfacial region is limited, we will for the present stick to simple models, hoping that they will give a qualitively correct representation of the physical system.

When a single ion approaches a dielectric discontinuity it will feel a force, due to the change in the dielectric environment. This interaction can be formally represented by induced image charges. The image interaction is clearly of long range character. It is a peculiar feature of the mean field PB approximation, where one is dealing with many charges, that for a symmetric situation like a planar sheet, the long range effects of the image interactions cance [25, 26]. However, there is a difference between media with different ε_r in the reference state of the ion. Since the PB treatment is based on the primitive model with a dielectric continuum, it is natural to estimate the self energy of a single ion in the medium using the Born energy, which, for a spherical ion of charge ze and radius r_i, is [27]

$$G_{\text{Born}} = \frac{z^2e^2}{8\pi\varepsilon_0}\frac{1}{r_i}\left(\frac{\varepsilon_r}{1} - 1\right). \tag{11}$$

In a PB description of the electrostatics of a system with a dielectric discontinuity, this leads to a step-function potential at the dielectric discontinuity with the magnitude given by the difference in electrostatic solvation energy of the ion in the two media. Using the Born model, this would be

$$\Delta G_{\text{Born}} = \frac{z^2 e^2}{8\pi \varepsilon_r \varepsilon_0} \frac{1}{r_i} \left\{ \frac{1}{\varepsilon_a} - \frac{1}{\varepsilon_w} \right\}. \tag{12}$$

This is, for example, the potential that excludes small hydrophilic ions from apolar regions.

The amphiphilic ion, being attracted to the apolar region by the hydrophobic effect, will experience a more complex local potential, and the notion of a sharp dielectric discontinuity is really inadequate

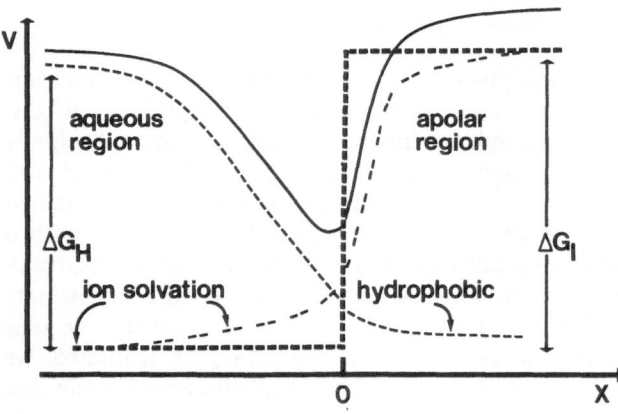

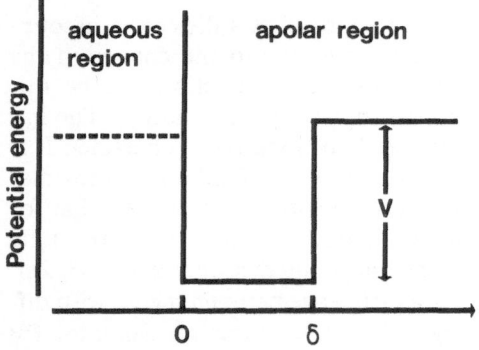

Fig. 3. A. Schematic illustration of the different interactions affecting an amphiphilic ion at the interface. ΔG_I is the difference in the solvation of the ionic group: (■ ■ ■) effective potential with sharp dielectric discontinuity; (------) the effective potential with a more gradual change in ε_r. ΔG_H is the change in hydrophobic energy in taking the apolar chain from a completely apolar to an aqueous environment. The full drawn line shows the total effective potential. B. Model potential used in the calculation. The aqueous region is treated as a separate pseudo-phase

when considering the situation in molecular detail. The potentials present at the interface are schematically illustrated in Fig. 3.

Having identified the main physical mechanism responsible for the special affinity of the amphiphilic ion to the interface we will now introduce a simple model representing these effects. The short range potentials at the interface are represented by a square well potential inside the apolar region of width δ and depth V, where the width should be of the order of 5 Å and the depth less than the difference in the solvation energy of Eq. (12) (see Fig. 3). In solving the PB equation the extra boundary condition is that at $x = \delta$. The potential is continuous while the ion concentration changes by a factor $\exp(-V/kT)$. This model has previously been used in describing specific ion binding effects [28] in lamellar liquid crystals and, in general, a numerical solution has to be used. In the present case, where the majority of ions are expected to be located in the region $0 < x < \delta$, one can obtain an approximate solution to the problem in a simple way. The ion concentration profile in the high affinity region at the surface can, to a reasonable approximation, be obtained by using the boundary condition that all ions are in this region. In Eq. (3) the parameter s is obtained by setting $b = \delta$. The solution in Eqs. (2, 4) applies with a relevant choice of constants and at the border of the square well potential the ion concentration is

$$c(\delta_-) = 2kT\varepsilon_a\varepsilon_0 (s_\delta/z_a e\delta)^2 \tag{13}$$

where s_δ is close to $\pi/2$.

Immediately outside the square well the concentration is

$$c(\delta_+) = \exp(-V/kT)c(\delta_-).$$

The small amount of ions n_{apolar} outside the square well can now be determined by using the contact value theorem [29, 30]

$$c(\delta_+) = (en_{\text{apolar}}/A)^2/(2kT\varepsilon_a\varepsilon_0) - c(a) . \tag{14}$$

Since, for cases of practical importance, $c(\delta_+) > c(a)$ (Eq. (14)) provides a good measure of n_{apolar}/A by setting $c(a) \simeq 0$ and

$$n_{\text{apolar}}/A \simeq \exp\{-V/2kT\}(kT\varepsilon_a\varepsilon_0 \pi/e^2\delta) \tag{15}$$

which in turn can be used in Eq. (3) to obtain the

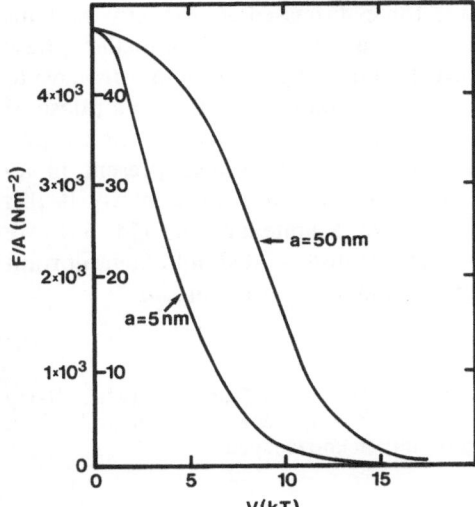

Fig. 4. The variation of the double layer force with the depth of the local potential V for an amphiphilic ion at the interface at two values for the thickness of the apolar lamellae. $2a = 10$ nm and $2a = 100$ nm. Left scale refers to $a = 5$ nm and right scale $a = 50$ nm. The van der Waals force is ~ 60 N m^{-2} at $a = 5$ nm and $\sim 6 \cdot 10^{-2}$ Nm^{-2} at $a = 50$ nm. Parameters $\varepsilon_a = 4$, $T = 300$, $\delta = 0.5$ nm

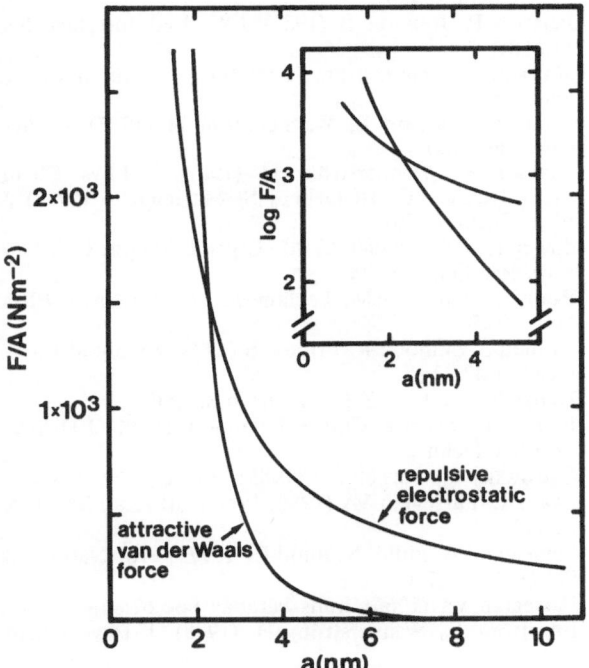

Fig. 5. Calculated distance dependence of the force. $V = 8$ kT; other parameters as in Fig. 4. For comparison, the van der Waals force calculated from Eq. (1) with $H = 6 \cdot 10^{-21}$ J is also shown. Insert shows the short range part with a logarithmic scale

boundary condition determining the ion distribution in the apolar region. Equations (8, 9) are still valid for the force between the lamellae, while in this case s_a can be expected to be substantially lower than $\pi/2$.

The effects of the variation of the calculated force or the depth of the potential well are illustrated in Fig. 4 for two different separations. Particularly at the large separation, the well has to be fairly deep before the double layer repulsion is weaker than the estimated van der Waals' force. In fact, at sufficiently large separations, the double layer force will always beat the van der Waals' force, irrespective of the value of V. In Fig. 5 the calculated distance dependence is shown for a value of 8 kT for V, showing a crossover between the double layer force and the van der Waals' force at small thicknesses.

These calculations show that one can obtain a substantial double layer repulsion also in the presence of a local affinity of the surfactant ion for the interface. In trying to estimate the importance of the effect, the crucial point is the depth of the potential well. An upper bound to the solvation energy effect is obtained from ΔG_{Born} in Eq. (12). The crucial parameter is the radius and the relevant value of ε_a. A small ΔG_{Born} can be obtained by a relatively high ε_a of the apolar region and by a large effective radius. The presence of substantial quantities of alcohol in the alkane tends to increase ε_a and a large radius can be obtained either by having the charge surrounded by an apolar shell, as in the tetrabutylammonium ion, or by having the charge delocalized, as in a benzene ring. For a compound like octylbenzenesulfonate both these mechanisms are to some extent in operation. In addition, one has to realize that in the Born model the medium is considered to be homogeneous, but the real system typically contains, in addition to the apolar alkane, substantial amounts of alcohol and, on a molar concentration basis, nearly as much water. There will clearly be a strong affinity for the water molecules to the first hydration shells of the ion leading to a stabilization of the ion [31]. A further factor tending to decrease V is illustrated in Fig. 3. The potential well is generated by the combined action of the solvation energy and the hydrophobic energy and the depth of the well is expected to be smaller than the change in electrostatic solvation energy between the two media. Combining all these considerations, it is certainly difficult to get a precise estimate of a realistic value for V. However, for the SOBS system studied by Larché et al. it should be in the range of 5 to 15 kT. For values in the upper range of this interval, the double layer forces are probably negligible, while in the lower range they dominate other forces.

Conclusions

The calculations presented above show that a double layer force can be generated across the apolar region of a lamellar liquid crystal in the presence of a suitable ionic amphiphile. Such a force can be responsible for the stability of the alkane swollen lamellar phase relative to a phase separation giving the isotropic apolar liquid. The basis for the existence of this force is the fact that the system is microscopically heterogeneous, so that the small hydrophilic counterion of the ionic surfactant is spacially separated from the amphiphilic ion, due to the combined effect of the solvation of the ions and the hydrophobic interaction.

The amphiphilic ion will have a preference for the interface between the polar and non-polar regions through two mechanisms. The long-range electrostatic forces act by themselves to give a high density of amphiphilic ions in the interfacial region. If this is the only mechanism considered, there will be a substantial double layer repulsion, as calculated from the PB equation. In fact, the electrostatic interactions could be so strong that fluctuations are important and an attractive layer force could appear [27, 32]. In the real system there will also be short range forces localizing the surfactant at the interface, due basically to the solvation of the ionic group. If these short range terms are very strong, not enough amphiphilic ions will be present in the apolar region to produce a sizeable double layer repulsion. By having a sufficiently lipophilic ion present these double layer forces can be induced. At present, it is not clear whether commonly used amphiphilic ions are lipophilic enough to generate this effect, but the results of Larché et al. [11] indicates that this might be the case.

Above we have examined in detail one contribution to the force between two surfactant monolayers across an apolar region. The total force is due to a combination of several effects. There is the van der Waals force (Eq. (1)) which tends to make the two monolayers form a simple bilayer. When an alkane is dissolved in such a bilayer there is an entropy due to the mixing with the alkane chains of the surfactant [10]. This results in a strong but short range repulsive force and it leads to a certain swelling [33]. For lecithin systems it seems that these two factors account for the behaviour relative to alkanes. In addition to the entropy of mixing there is also an entropy associated with the long wavelength undulations of the aqueous layers which give a long-range repulsive force [15]. To obtain a complete description of the swelling of a lamellar phase in oil all these contributions to the force have to be considered.

The repulsive force discussed above only explains the stability of the alkane swollen lamellar phase relative to a neat liquid. Other considerations have to be made to explain the stability relative to a phase of the microemulsion type. Conversely, if the force is present in the lamellar phase it is also present in an analogous microemulsion phase, particularly if this contains aggregates of a lamellar type [34, 35]. We also note that the force discussed should be important for the stability of water in oil emulsions.

Acknowledgments

Illuminating discussions with Francis Larché, Barry Ninham, Hugo Christensson, Bengt Jönsson and Björn Lindman are gratefully acknowledged.

References

1. Gulik-Krzywicki T, Tardieu A, Luzzati V (1969) Mol Cryst Liq Cryst 8:285
2. Cluny JS, Goodman JF, Ingram BT (1971) Surf Colloid Sci 3:167
3. Ekwall P (1965) in: Brown GH (ed) Advances in Liquid Crystals. Academic Press, New York, vol 1:1
4. Tiddy GJT (1980) Phys Rep 57:1
5. Rydhag L, Gabran T (1982) Chem Phys Lipids 30:309
6. Fontell K, Ceglie A, Lindman B, Ninham B (in press) Acta Chem Scand
7. Persson P, Jönsson B (1987) J Colloid Interface Sci 115:507
8. Jönsson B, Wennerström H (1981) J Colloid Interface Sci 80:482
9. Khan A, Jönsson B, Wennerström H (1985) J Phys Chem 89:5180
10. Jönsson B, Wennerström H (1987) J Phys Chem 91:338Larché FC, El Qebbaj S, Marignan J (1986) J Phys Chem 90:707
12. diMeglio JM, Dvolaitzky M, Leger L Taupin C (1985) Phys Rev Lett 54:1696
13. Roux D, Bellocq AM, Leblanc MS (1983) Chem Phys Lett 94:156
14. Ahmad S, Shinoda K, Friberg S (1974) J Colloid Interface Sci 47:32
15. Helfrich W (1978) Z Naturforsch 33A:305
16. Biais J, Bothorel P, Clin B, Lalanne P (1981) J Dispersion Sci Techn 2:67
17. Ostrowsky N, Sornette D (1985) D Chem Scr 25:108
18. Evans E, Parsegian VA (1986) Proc Natl Acad Sci USA 83:7132
19. Parsegian VA, Fuller N, Rand RP (1979) Proc Natl Acad Sci USA 76:2750
20. Parsegian VA (1966) Trans Faraday Soc 62:848
21. Engström S, Wennerström H (1978) J Phys Chem 82:2711
22. Lindman B, Lindblom G, Wennerström H, Gustavsson H (1977) In: Mittal KL (ed) Micellization, Solubilization and Microemulsions. Plenum Press, New York, vol 1:195
23. Wennerström H, Lindman B, Lindblom G, Tiddy GJT (1979) J Chem Soc Faraday Trans I 75:663

24. Marcus RA (1955) J Chem Phys 23:1057
25. Jönsson B, Wennerström H, Halle B (1980) J Phys Chem 84:2179
26. Bratko D, Jönsson B, Wennerström H (1986) Chem Phys Lett 128:449
27. Israelachvili, J (1985) Intermolecular and Surface Forces, Academic Press, New York
28. Söderman O, Engström S, Wennerström H (1980) J Colloid Interface Sci 78:110
29. Henderson D, Blum L (1978) J Chem Phys 69:5441
30. Wennerström H, Jönsson B, Linse P (1982) J Chem Phys 76:4665
31. McDonald RC (1976) Biochim Biophys Acta 448:193
32. Guldbrand L, Jönsson B, Wennerström H, Linse P (1984) J Chem Phys 80:2221
33. Gruen DWR, Haydon DA (1980) Pure Appl Chem 52:1229
34. Shinoda K (1983) Progr Colloid Polym Sci 68:3
35. Olsson U, Shinoda K, Lindman B (1986) J Phys Chem 90:4083

Received October 28, 1986;
accepted November 28, 1986

Author's address:

Håkan Wennerström
Division Physical Chemistry 1
Chemical Center
P. O. Box 124
S-22100 Lund, Sweden

Progress in Colloid & Polymer Science   Progr Colloid & Polymer Sci 74:38—47 (1987)

A theoretical study of the effect of long-chain aliphatic alcohols on the stability of surfactant micelles

S. Ljunggren and J. C. Eriksson

Department of Physical Chemistry, The Royal Institute of Technology, Stockholm, Sweden

Abstract: Employing the theoretical approach and computational methods described previously [9], calculations have been made on the two-dimensional size/composition distribution of spherical micelles composed of sodium dodecylsulphate (SDS) and dodecylalcohol (DOH) monomers. These calculations show fair agreement with experimental CMC and solubilization capacity results. For rod-shaped micelles our calculations indicate that the critical condition for the formation of rods, i.e. $\beta = 0$, where βkT is the difference in molecular free energy per SDS/DOH unit between the middle part of the micelle and the solution, is approached at significantly lower salt concentrations upon raising the DOH/SDS ratio.

The incorporation of DOH in the SDS micelles is driven by the free energy of mixing the monomers within the micelles and by the reduction obtained of their electrostatic free energy. Since a slight swelling is noted, there is a counteracting increase in chain conformational free energy. Moreover, the hydrophobic driving force of aggregation is considerably weaker for DOH than for SDS because the monomer concentration of DOH is comparatively low.

Key words: Molecular theory of micelles, CMC-lowering, sphere-rod micelle transition, energetics of surfactant, alcohol mixing in micelles

Introduction

The addition of long-chain aliphatic alcohols usually has a rather significant influence on surfactant aggregation. In particular, it promotes the formation of rod-shaped micelles [1] and lamellar liquid-crystalline phases [2]. The stabilizing effect of aliphatic alcohols has been treated theoretically during recent years by Jönsson and Wennerström, in the course of their work on ternary H_2O/surfactant/long-chain alcohol phase diagrams [3] and by Rao and Ruckenstein [4]. However, these calculations involve certain simplifying assumptions which need to be further elucidated in order to reach a more satisfactory description at the molecular level. One such assumption, made by Jönsson and Wennerström (loc. cit.), is that ordinary spherical surfactant micelles can be treated as being effectively monodisperse. This is in obvious disagreement both with experimental results on the width of the micellar size distribution [5, 6] and with previous theoretical calculations [7, 8, 9]. Another assumption of this kind which is implicitly introduced in the Jönsson-Wennerström and Rao-Ruckenstein schemes is that the conformational free energy of packing the

chains into hydrocarbon cores is size- and shape-independent. Although this particular free energy contribution is of a comparatively small magnitude it varies at an appreciable rate with the core radius R. In fact, from our previous investigations of pure soap micelles [9, 10, 11] we may conclude that a consistent theoretical description of micellar aggregates seems feasible only when the conformational free energy of the hydrocarbon chains in the interior of a micelle is included in the theory. Approximate values of this free energy are available for different geometries through the calculations of Gruen et al. [12, 13] which are based on a mean-field, single-chain approach.

There are as yet no detailed statistical-mechanical calculations of the conformational free energy of the hydrocarbon core of a mixed micelle composed of hydrocarbon chains of *different* lengths. Hence the only case that is at present amenable to a complete theoretical treatment is when the hydrocarbon chains of the surfactant and of the alcohol are identical, i.e. have the same lengths. In view of this we have chosen to treat that particular case as a pilot case, to study the effects of long-chain aliphatic alcohols on equilibrium ensembles of spherical and rod-shaped surfactant micelles. More specifically, the calculations presented below refer to an ionic C_{12} surfactant, like sodium dodecylsulphate (SDS), and n-dodecyl alcohol (DOH). However, the theory may be extended to obtain a rather more approximate treatment even of the case where the chain length of the alcohol is somewhat less than that of the surfactant, as we will also shortly discuss in the present paper.

Spherical SDS/DOH micelles

Let us introduce the Gibbs free energy excess of a single micelle defined as

$$G = F + p_e V - \mu_1 n_1 - \mu_2^+ n_2^+ - \mu_2^- n_2^- . \tag{1}$$

This Gibbs free energy function involves the external pressure, p_e, that is supposed to act on the hydrocar-

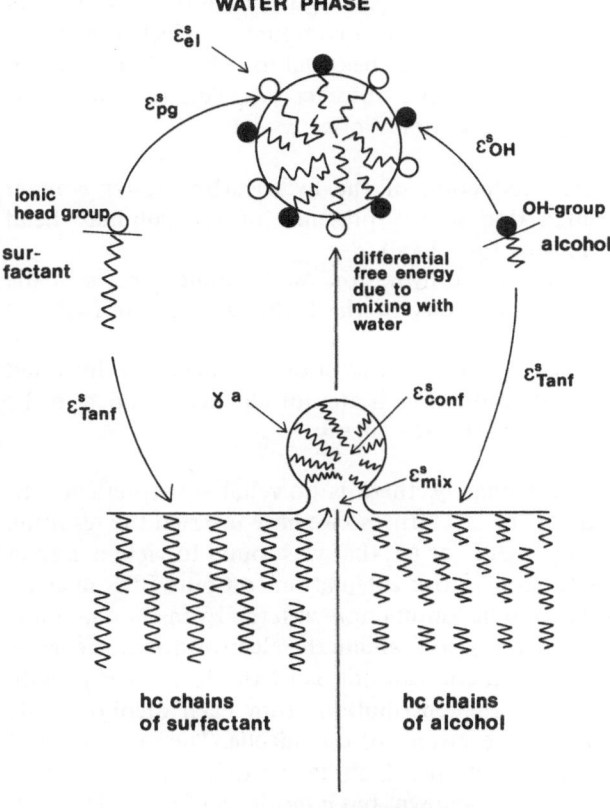

Fig. 1. Drawing illustrating the various contributions to the free energy of forming a micelle composed of N_2 surfactant monomers and N_3 long-chain aliphatic alcohol monomers. The hydrocarbon core is assembled separately (lower part) and dispersed in the water phase. The head groups are added onto the core directly from the water phase

bon core volume, V, of the micelle. According to Eq. (1) the free energy associated with the *excess* of water (n_1), counterions (n_2^+) and coions (n_2^-) is deducted from $F + p_e V$ where F is the net Helmholtz free energy of the micelle, employing the hydrocarbon core radius R as the dividing surface. Since we shall limit our treatment to solutions which are dilute in micelles, the chemical potentials μ_2^+ and μ_2^- of the cations and surfactant anions are given by the usual expressions

$$\mu_2^+ = \mu_2^{+0} + kT \ln c_2^+$$
$$\mu_2^- = \mu_2^{-0} + kT \ln c_2^- \tag{2}$$

where $c_2^+ = c_2^- = c_2$ are the ion concentrations in the (undisturbed) solution of monomers.

The fundamental quantity that determines the distribution of micelles with different numbers, N_2 and N_3, of surfactant and alcohol monomers, respectively is [9]

$$\varepsilon(N_2, N_3) = G(N_2, N_3) - N_2 \mu_2^- - N_3 \mu_3 \tag{3}$$

where μ_2^- and μ_3 are the chemical potentials of the surfactant and alcohol monomers in solution. ε is thus the free energy of forming one additional micelle in a solution where there is already equilibrium between micelles and monomers. In the following, all free energies of this kind will be given in kT units. The different contributions to ε are illustrated in Fig. 1, where the bulk hydrocarbon phases serve as purely hypothetical intermediates.

The first and most well-known contribution to $\varepsilon^s(N_2, N_3)$ of a spherical SDS/DOH micelle is the free energy, as quantified by Tanford [14], required to transfer N_2 hydrocarbon chains at the mole fraction x_2^- in water and N_3 chains at the mole fractions x_3 in water to the corresponding pure bulk hydrocarbons which will serve as reference states. This contribution is of the following magnitude (for the whole micelle):

$$\varepsilon_{Tanf}^s = -[19.960 \, (N_2 + N_3) + N_2 \ln x_2^- + N_3 \ln x_3]. \tag{4}$$

The second free energy contribution is that caused by the residual hydrocarbon/water interface, $\gamma_{hc/w} A / kT$, $\gamma_{hc/w}$ denoting the hydrocarbon/water interfacial tension and, A, the overall area of the hydrocarbon core.

The third contribution is the electrostatic one which can be calculated with sufficient accuracy for the present purpose using a simple formula suggested by Evans and Ninham [15]:

$$N_2 \varepsilon_{el}^s = 2N_2 [\ln(S + \sqrt{S^2+1}) - (\sqrt{S^2+1} - 1)/S]$$
$$- (4N_2/\kappa R_{el} S) \ln[(1 + \sqrt{S^2+1})/2] \qquad (5)$$

where ε_{el}^s is the contribution per charged head group and R_{el} is the radius through the centre of the charges which is some 3 Å larger than

$$R = [3(N_2 + N_3)\hat{V}/4\pi]^{1/3},$$

$\hat{V}$ denoting the volume of one hydrocarbon chain, and where the parameters S and κ are:

$$S = \sigma(8c_s \varepsilon_o \varepsilon_r RT)^{-1/2} \qquad (6)$$

$$\kappa = [2c_s N_A^2 e^2 / \varepsilon_o \varepsilon_r RT]^{1/2} \qquad (7)$$

R denoting the gas constant and N_A the Avogadro number. Here, c_s is the total salt concentration (including c_2^-) and

$$\sigma = e/a \qquad (8)$$

the surface charge density, e being the charge of the surfactant head group and a the corresponding surface area, i.e.

$$a = 4\pi R^2 / N_2. \qquad (9)$$

The fourth contribution to $\varepsilon^s(N_2, N_3)$ is the hydrocarbon chain conformational free energy inside the micelle. From the curves published by Gruen and de Lacey [12] this contribution may be set approximately equal to

$$(N_2 + N_3)\varepsilon_{conf}^s = (N_2 + N_3)(7.612 - 0.9272 \times 10^{10} R$$
$$+ 0.03233 \times 10^{20} R^2$$
$$- 6.688 \times 10^{25} R^3) . \qquad (10)$$

The inclusion of ε_{conf}^s in the expression for $\varepsilon^s(N_2, N_3)$ is essential since, together with ε_{el}^s, it counteracts the fall-off of $a\gamma_{hc/w}$ with increasing values of N_2 owing to its rapid variation with the radius, R, in such a way that a minimum of $\varepsilon^s(N_2, N_3)$ is generated in the right micelle size range (cf. Ref. [9]). Previous experience with pure surfactant micelles (without alcohol) shows that by including ε_{conf}^s it is possible to account for both the stability of the normal micelles and the sparsity of very small micelles with $N_2 \approx 10-20$ (as required by the Aniansson-Wall theory of micellization kinetics [6]), employing the macroscopic value $\gamma_{hc/w} = 50$ mJ/m.

The above formula for ε_{conf}^s is rather approximate. In a later publication by Gruen [13], another formula was presented which yields smaller values for ε_{conf}^s. In order to retain the balance against the hydrocarbon/water contact free energy, $a\gamma_{hc/w}/kT$, a smaller value of $\gamma_{hc/w}$ would then have to be used, about 40 mJ/m^2. This is, indeed, quite feasible since it is natural to expect a certain curvature dependence of the interfacial tension $\gamma_{hc/w}$, although so far, there is no theory for this kind of effect. Our previous calculations on pure surfactant micelles indicate that this set of parameters might, in fact, be preferable since $\gamma_{hc/w} = 50$ mJ/m^2 yields a size distribution curve which appears to be too narrow (cf. Ref. [9]). However, in order to preserve conformity with our earlier calculations, the values of $\gamma_{hc/w}$ and ε_{conf}^s used previously were also employed throughout the present calculations.

The fifth contribution to $\varepsilon^s(N_2, N_3)$ is $N_2 \varepsilon_{pg}^s$, i.e. a constant contribution per surfactant head group which earlier, when making use of Jönsson's Poisson-Boltzmann program to compute the electrostatic free energy, was found to be equal to -1.310 in the case of pure spherical SDS micelles [9]. This term accounts for several additional effects, viz:

1. The reduction of the hydrocarbon/water contact area due to the presence of the sulphate head groups ($\approx -3\,kT$),
2. Asymmetric hydration of the ionic species in the close proximity of the hydrocarbon core ($\approx 1\,kT$) and
3. Repulsive local interactions between the hydrated head groups which are not already included in the electrostatic free energy.

Unfortunately, there is no reliable theoretical estimate of the last term, so we have inserted the resulting value -1.385 for ε_{pg}^s that was found to yield a correct CMC value in our calculations on pure SDS micelles when used in conjunction with the Evans-Ninham formula, Eq. (5), to evaluate the electrostatic free energy.

The sixth contribution is of the form $N_3 \varepsilon_{OH}^s$, with ε_{OH}^s being a contribution from each alcoholic OH group at the surface of the micelle. The magnitude of this contribution, which has an origin similar to that of ε_{pg}^s, is not known, but considering factors 1) and 2) above and the fact that OH is a rather small head group, it is not unreasonable to put ε_{OH}^s equal to zero as we will tentatively do below. Of course, the form of the fifth and sixth contributions presumes that these contributions are linearly dependent on N_2 and N_3 which in itself is an approximation.

By adding up the different contributions and also inserting the free energy of mixing the surfactant and alcohol monomers in the micelle we arrive at the following expression for $\varepsilon^s(N_2, N_3)$:

$$\varepsilon^s(N_2, N_3) = -19.960(N_2+N_3) - N_2 \ln x_2^- - N_3 \ln x_3$$
$$+ (N_2+N_3)\varepsilon_{conf}^s + N_2\varepsilon_{el}^s + \gamma_{hc/w}N_2 a/kT$$
$$+ N_2\varepsilon_{pg}^s + N_3\varepsilon_{OH}^s + \ln \begin{pmatrix} N_2+N_3 \\ N_3 \end{pmatrix} \quad (11)$$

where the last term is the combinatorial expression for the free energy of mixing. For spherical micelles with low N_3 values, it may differ by as much as $2\,kT$ from the usual expression $kT\sum_i x_i \ln x_i$.

Collecting the unknown parameters in a constant term Q^s:

$$Q^s = \varepsilon_{OH}^s - \ln x_3 \quad (12)$$

and introducing the function

$$W^s(N_2, N_3) = -19.960(N_2+N_3) - N_2 \ln x_2^-$$
$$+ (N_2+N_3)\varepsilon_{conf}^s + N_2\varepsilon_{el}^s + \gamma_{hc/w}N_2 a/kT$$
$$+ N_2\varepsilon_{pg}^s + \ln \begin{pmatrix} N_2+N_3 \\ N_3 \end{pmatrix} \quad (13)$$

our expression for $\varepsilon^s(N_2, N_3)$ may be written as

$$\varepsilon^s(N_2, N_3) = W^s(N_2, N_3) + N_3 Q^s. \quad (14)$$

Glancing at Eq. (3) it might appear that $\varepsilon^s(N_2, N_3)$ should be equal to zero at equilibrium. There is, however, an additional free energy contribution arising from the dispersion of the micelles in the solution, so that the true equilibrium condition takes on the following form

$$\varepsilon^s(N_2, N_3) + \ln \phi(N_2, N_3) = 0 \quad (15)$$

remembering that $\varepsilon^s(N_2, N_3)$ is supposed to be given in kT-units. Hence,

$$\phi(N_2, N_3) = \exp\left[-\varepsilon^s(N_2, N_3)\right] \quad (16)$$

where $\phi(N_2, N_3)$ is a suitable concentration measure.

It has often been assumed previously that the mole fraction $x(N_2, N_3)$ of micelles would be the correct concentration measure in this context. However, we believe that $x(N_2, N_3)$ should rather be replaced by the volume fraction $\phi(N_2, N_3)$, as can be shown by the following simplified argument.

Referring to a lattice model, the integral entropy of mixing large, compact molecules, B, with small molecules, A, is, at low concentrations of B, approximately

$$\Delta S_{mix}/k = -N_B \ln \phi_B + N_B b/2v_B^0 \quad (17)$$

where ϕ_B is the volume fraction of B, N_B the number of B particles, b the excluded volume in the lattice caused by a B particle ($\approx v_B^0$) and v_B^0 = the volume of a B particle.

However, since we are interested in the free energy change caused by adding one more micelle, we should actually calculate $\partial\Delta G_{mix}/\partial N_B$ where $\Delta G_{mix} = -T\Delta S_{mix}$. We thus obtain (in kT units)

$$\Delta\mu_B = \partial\Delta G_{mix}/\partial N_B \approx \ln \phi_B + 1/2\phi_A(1+\phi_B) \quad (18)$$

where we shall henceforth neglect the second term which is of the order of $1/2$. Thus, the free energy of mixing one micelle with the water becomes $kT \ln \phi(N_2, N_3)$, ϕ denoting the volume fraction of micelles.

It is highly dubious whether there is any communal entropy associated with the subdivision of the (hydrocarbon) bulk phase into micellar packages since the exchange of monomers between different micelles may not be much slower than between different parts of the bulk phase. In any case it is easy to show that this contribution would be of a totally negligible magnitude (ca. one twentieth of a kT-unit).

In addition to Eqs. (15) and (16), there is another way of looking upon the distribution of micelles with different overall aggregation numbers and composition in the solution. The intrinsic chemical potentials of surfactant and alcohol in a micelle with the aggregation numbers N_2 and N_3 can be defined by the following partial derivatives

$$[\partial G(N_2, N_3)/\partial N_2]_{T, p_e, N_3} = \mu_{2, mic}^- \quad (19)$$

$$[\partial G(N_2, N_3)/\partial N_3]_{T, p_e, N_2} = \mu_{3, mic} \quad (20)$$

Thus

$$[\partial\varepsilon^s(N_2, N_3)/\partial N_2]_{T, P_e, N_3, \mu_2^-, \mu_3} = (\mu_{2, mic}^- - \mu_2^-)/kT \quad (21)$$

$$[\partial\varepsilon^s(N_2, N_3)/\partial N_3]_{T, p_e, N_3, \mu_2^-, \mu_3} = (\mu_{3, mic} - \mu_3)/kT \quad (22)$$

In a solution with fixed monomer concentrations of surfactant and alcohol the expressions in Eqs. (21) and (22) can be zero only for micelles with a definite pair of values of N_2 and N_3, corresponding to the peak value of the function $\phi(N_2, N_3)$. This is *the equilibrium micelle*. The existence of micelles with values of N_2 and N_3 different from those of the equilibrium micelle can thus be regarded as a *fluctuation phenomenon* which is, by the way, to be expected, since the micelles are small systems composed of a rather limited number of molecules. Thus the micelle size fluctuates around the radius of the equilibrium micelle and, moreover, the composition of a mixed micelle is subject to significant fluctuations as emerges out of the present calculations.

In order to solve for the equilibrium micelle we set the expressions in Eqs. (21) and (22) equal to zero. With the use of Eq. (14) this becomes

$$[\partial W^s(N_2, N_3)/\partial N_2]_{N_3} = 0 \qquad (23)$$

$$[\partial W^s(N_2, N_3)/\partial N_3]_{N_2} + Q^s = 0 \qquad (24)$$

Employing N_3 as the independent variable, the above Eqs. (23) and (24) can be used to determine N_2 for the equilibrium micelle and, in addition, the initially unknown constant Q^s. Then $\varepsilon^s(N_2, N_3)$ and $\phi(N_2, N_3)$ can easily be calculated from Eqs. (14) and (16) for different values of N_2 and N_3, which yields the distribution of micelles in the N_2- and N_3-plane (Fig. 2).

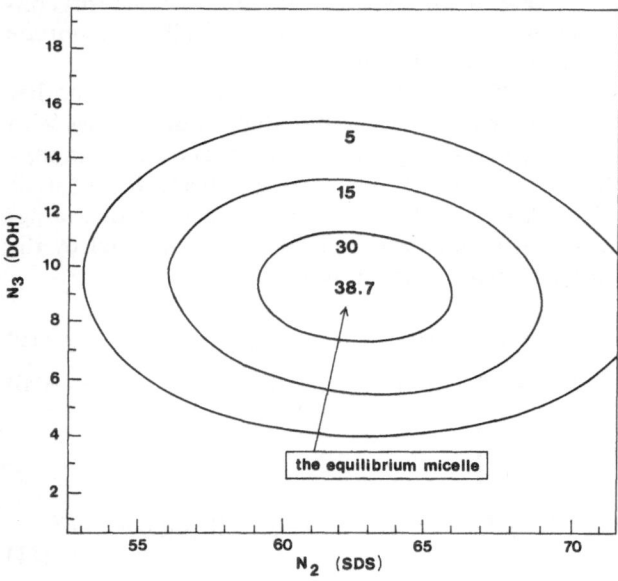

Fig. 2. Example of a two-dimensional micelle size distribution generated at $c_2^- = 7.64$ mM and $x_{OH}^{tot} = 2.18\%$ which involves fluctuations in the numbers of aggregated SDS and DOH monomers around the equilibrium micelle with $N_2 = 62$ and $N_3 = 9$. The parameter values on the curves are $\phi(N_2, N_3) \times 10^7$

The width of the distribution does not seem to depend very much on the concentration of micelles. Note also that the N_3-fluctuations are large, relatively speaking. By summing over N_2 and N_3 the total micelle volume fraction is obtained

$$\phi_{tot} = \sum_{N_2, N_3} \phi(N_2, N_3) \qquad (25)$$

as well as the (number) average values of N_2 and N_3:

$$\langle N_2 \rangle = \frac{\sum\limits_{N_2, N_3} N_2 \phi(N_2, N_3)/(N_2 + N_3)}{\sum\limits_{N_2, N_3} \phi(N_2, N_3)/(N_2 + N_3)} \qquad (26)$$

$$\langle N_3 \rangle = \frac{\sum\limits_{N_2, N_3} N_3 \phi(N_2, N_3)/(N_2 + N_3)}{\sum\limits_{N_2, N_3} \phi(N_2, N_3)/(N_2 + N_3)} \qquad (27)$$

This yields the average molar fraction of alcohol in the micelles

$$x_{OH}^{mic} = \langle N_2 \rangle/(\langle N_2 \rangle + \langle N_3 \rangle) \ . \qquad (28)$$

The molar fraction x^{tot} in the whole solution was obtained as

$$x_{OH}^{tot} = \frac{x_{OH}^{mic} \phi_{tot}/v + c_3}{\phi_{tot}/v + c_3 + c_2^-} \qquad (29)$$

where v is the molar volume of the hydrocarbon chain (set equal to 0.21138 dm^3/mol) and where c_3 is the concentration of alcohol in monomeric form in the solution.

The following approximate, linear relation was found to hold for the equilibrium micelle

$$N_2 = 68 - 0.65 N_3 \qquad (30)$$

almost independently of the total concentration of micelles. This implies that the addition of DOH causes only a rather restricted growth of the equilibrium micelle. As a first (crude) approximation one might even claim that one OH group *replaces* one sulphate group in the micelle surface.

The value of ε_{OH}^s can, in principle, be obtained from experiment. Unfortunately, the relevant information is lacking as regards the exact value of ϕ_{tot} which was chosen to define the CMC in the experimental studies reported so far and this prevents us from attempting such an approach. With $\varepsilon^s_{OH} = 0$ the values shown in Table 1 were obtained for c^-_2 at the concentration $\phi_{tot} = 3 \times 10^{-4}$ which is assumed here to correspond to the CMC. The values included yield a distribution coefficient $K = x^{mic}_{OH}/vc_3$ which varies between 110461 at $x_3 = 1.0 \times 10^{-7}$ and 80345 at $x_3 = 3.0 \times 10^{-7}$ (the latter value being hypothetical because it exceeds the solubility of the alcohol in water) (Fig. 3).

Table 1. Calculation results obtained at the fixed total volume fraction $\phi_{tot} = 3.0 \times 10^{-4}$ of spherical SDS/DOH micelles (that is assumed to define the CMC) upon varying the DOH content and assuming $\varepsilon^s_{OH} = 0$

c^-_2 (mM)	x_3	x^{mic}_{OH} (%)	x^{tot}_{OH} (%)	$c^-_{2,tot}$ (mM)
8.348	0	0	0	9.767
7.644	1.0×10^{-7}	12.96	2.165	8.879
7.040	2.0×10^{-7}	21.80	3.800	8.150
6.510	3.0×10^{-7}	28.28	5.300	7.528

The literature values of the CMC of SDS containing variable amounts of added DOH were determined by Miura and Arichi in the following way [16]. A weighted amount of DOH was added to 250 ml of 0.03 M SDS solution. The amount of DOH added was always below saturation. Regarding the method of determining the CMC of the solution it was only stated that the conductivity of the solution was measured. Owing to the sparsity of information, it is very difficult to assess to what exact value of the micelle concentration ϕ_{tot} the reported CMC data refer. Since the ratio of x^{tot}_{OH} to x^{mic}_{OH} varies rapidly at low ϕ_{tot}, the value of ϕ_{tot} chosen as a criterion for the CMC has a strong influence on the course of the CMC versus x^{tot}_{OH} curve.

The reason for our choice of such a comparatively high value of ϕ_{tot} as 3×10^{-4} at the CMC is the very low solubility of DOH in pure water ($x_3 = 2.449 \times 10^{-7}$). This would make it impossible to obtain the high values of the molar ratio DOH/SDS reported in the experimental studies unless ϕ_{tot} at the CMC had actually been chosen rather high. Fig. 4 shows a comparison between the experimental results

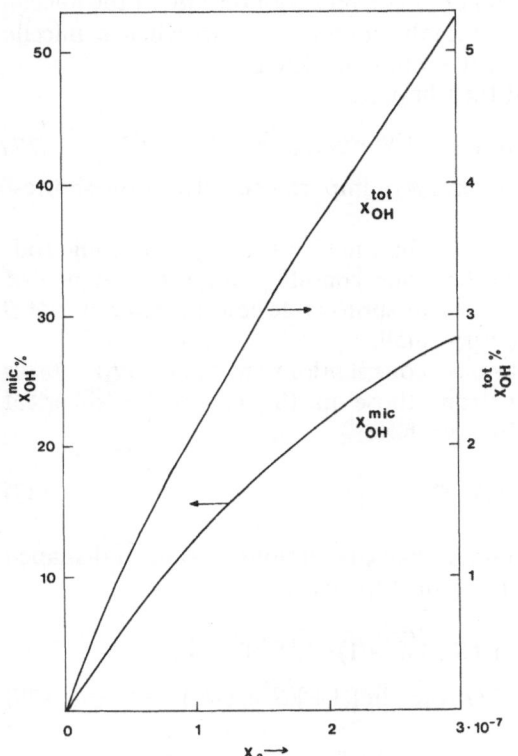

Fig. 3. The average mole fraction of DOH in the SDS/DOH micelle, x^{mic}_{OH}, and the overall mole fraction of DOH (excluding the water), x^{tot}_{OH}, in micellar or monomeric form plotted versus the monomer mole fraction x_3 of DOH in the solution

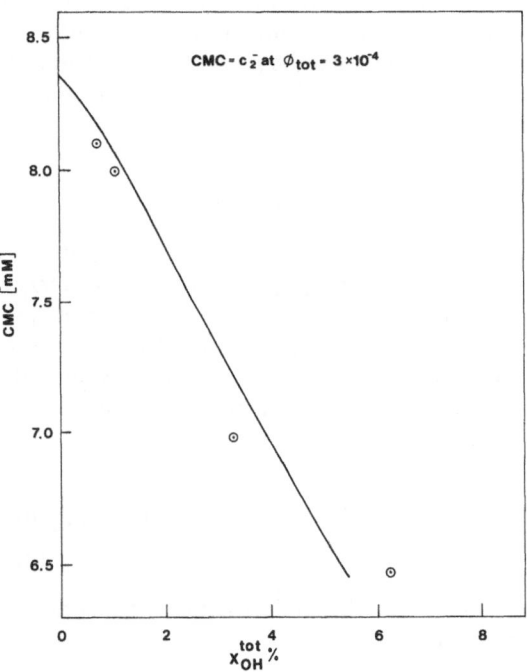

Fig. 4. Comparison between the experimental CMC results of Miura and Arichi [16] and our calculations as regards the effect of adding DOH to a SDS solution. The CMC was defined in the calculations as the monomer concentration c^-_2 needed to yield the volume fraction $\phi_{tot} = 3.0 \times 10^{-4}$ of micelles

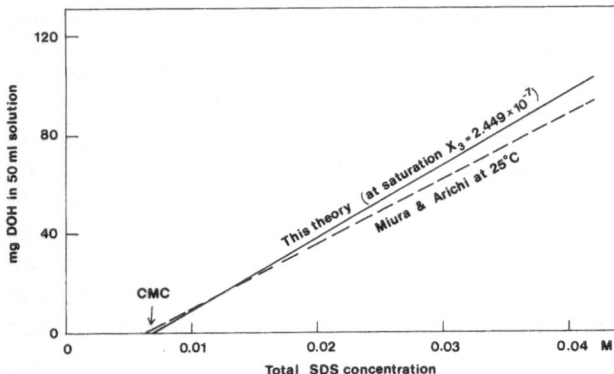

Fig. 5. Solubilization of DOH in SDS solutions. The total amount of dissolved DOH was determined at saturation with DOH

and our calculations, from which it appears that our theory can reproduce quite well the decrease of the CMC of SDS caused by the addition of DOH.

Miura and Arichi [16] also reported data on the maximum number of moles of DOH solubilized in SDS solutions of different concentrations (Fig. 5). The perfect straight line behaviour of the experimental curve (broken line) indicates that the composition of the solid phase does not change in this concentration interval. Assuming DOH to be present at its saturation concentration and setting $\varepsilon_{OH}^{s} = 0$ our model calculations yielded the solid line in Fig. 5.

Rod-shaped SDS/DOH micelles

In this case we are dealing with a broad micelle size distribution which is largely geared by the free energy parameter β that measures the work needed per monomer to form the central part of a long, rod-shaped micelle [10]. It is then convenient to subdivide the expression for $\varepsilon^{c}(N_2, N_3)$ in the following way

$$\varepsilon^{c}(N_2, N_3) = G(N_2, N_3) - (N_2 + N_3)\hat{\mu}_{\infty}^{c}$$
$$+ (N_2 + N_3)[\hat{\mu}_{\infty}^{c} - (1 - x_{OH})\mu_2^{-} - x_{OH}\mu_3]$$

$$(32)$$

where x_{OH} stands for the mole fraction of alcohol in the single micelle:

$$x_{OH} = N_3/(N_2 + N_3) \ . \tag{33}$$

In Eq. (32), $\hat{\mu}_{\infty}^{c}$ is the free energy per mixed monomer unit (of alcohol + surfactant) of an infinite rod-shaped micelle.

Let us now set

$$a + \delta = G(N_2, N_3) - (N_2, + N_3)\hat{\mu}_{\infty}^{c} \tag{34}$$

and

$$\beta = \hat{\mu}_{\infty}^{c} - (1 - x_{OH})\mu_2^{-} - x_{OH}\mu_3 \tag{35}$$

where

$$a = \lim_{N_2, N_3 \to \infty} [G(N_2, N_3) - (N_2 + N_3)\hat{\mu}_{\infty}^{c}] \ . \tag{36}$$

As before, we assume that the new free energy quantities a, β, δ defined here are given in kT-units. Physically, a corresponds to the work of formation of the end-caps of a rod-shaped micelle from the (infinite) cylindrical part of the micelle. For short, rod-shaped micelles a correction term $\delta(N_2, N_3)$ will have to be added since the transition zones between the two end-caps and the cylindrical part of the micelle will then tend to overlap to a greater or lesser extent.

It should be pointed out that a is considered as a constant and does not vary with the size of the micelle and that δ rapidly approaches zero when a micelle grows from the minimum length.

We will then have

$$\Phi(N_2, N_3) = e^{-\varepsilon^{c}(N_3, N_3)} = e^{-a - \delta - (N_2 + N_3)\beta} \tag{37}$$

where the δ-term is of importance only for the shortest rods.

Since $N_2 + N_3$ becomes very large for long rod-shaped micelles, one condition for the existence of such micelles at an appreciable concentration is that β should be very small.

The various contributions to $\varepsilon^{c}(N_2, N_3)$, differ somewhat from those in the case of a spherical micelle. Thus we have,

$$\varepsilon_{pg}^{c} = -0.5897 \tag{38}$$

as was obtained from calculations on pure, rod-shaped SDS-micelles [10]. Moreover,

$$\varepsilon_{el}^{c} = 2 [\ln(S + \sqrt{S^2 + 1}) - (\sqrt{S^2 + 1} - 1)/S]$$
$$- (2/\kappa R_{el} S) \ln [(1 + \sqrt{S^2 + 1})/2] \tag{39}$$

and with $R_m = R - 11.04$:

$$\varepsilon_{conf}^{c} = 0.555 + 2.288 \times 10^{-3} R_m + 2.404 \times 10^{-2} R_m^2$$
$$- 4.482 \times 10^{-3} R_m^3 - 1.482 \times 10^{-4} R_m^4$$
$$+ 1.902 \times 10^{-4} R_m^5 + 2.112 \times 10^{-5} R_m^6 \tag{40}$$

while $\varepsilon^c_{\text{Tanf}}$ is the same as $\varepsilon^s_{\text{Tanf}}$. In addition, we have the free energy of mixing the sulphate and OH monomers. Here the conventional expression $\Sigma x_i \ln x_i$ can be used since the micelles are large enough as systems.

If we now introduce the function

$$V^c(R, x_{\text{OH}}, c_s) = -19.960 + \varepsilon_{\text{conf}}(R)$$
$$+ (1 - x_{\text{OH}}) \varepsilon^c_{\text{el}}(a^c, c_s)$$
$$+ (1 - x_{\text{OH}}) a^c \gamma_{hc/w}/kT$$
$$+ (1 - x_{\text{OH}}) \varepsilon^c_{pg} + x_{\text{OH}} \ln(x_{\text{OH}})$$
$$+ (1 - x_{\text{OH}}) \ln(1 - x_{\text{OH}}) \qquad (41)$$

where c_s is to total salt concentration, the upper index c refers to the case of an infinite cylindrical micelle, and where the surface area per unit surface charge (i.e. per surfactant head group) is $a^c = 2\hat{V}/R$ ($\hat{V}$ is the

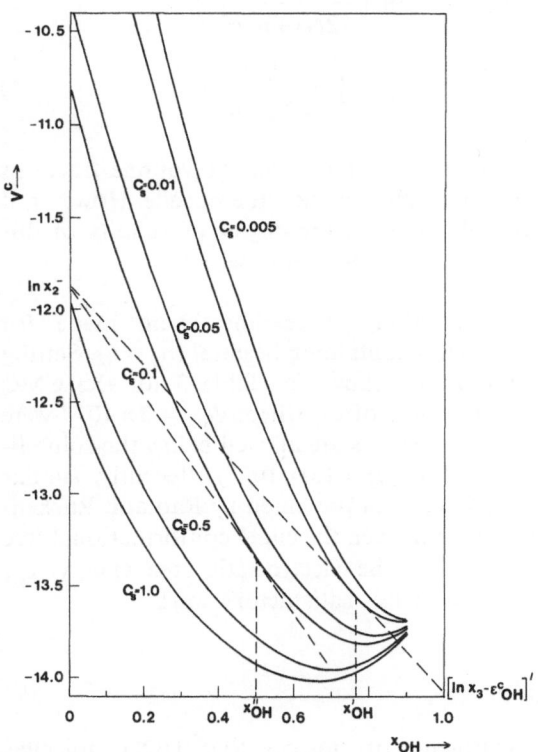

Fig. 6. Plots of the V^c-function for rod-shaped SDS/DOH micelles with the salt concentration c_s as parameter. The intercepts of a tangent on the vertical axes at $x_{\text{OH}} = 0$ and $x_{\text{OH}} = 1.0$ yield $\ln x^-_2$ and $\ln x_3 - \varepsilon^c_{\text{OH}}$. The dotted lines indicate that the critical condition, $\beta = 0$, for formation of rod-shaped micelles can be reached at a much lower salt concentration ($c_s = 0.01$ instead of 0.1) upon increasing x_{OH} from 0.5 to 0.77

volume per hydrocarbon chain), then β can be written in the following convenient way

$$\beta = V^c(R_{\min}, x_{\text{OH}}, c_s) - (1 - x_{\text{OH}}) \ln x^-_2 + x_{\text{OH}} Q^c \quad (42)$$

where

$$Q^c = \varepsilon^c_{\text{OH}} - \ln x_3 \qquad (43)$$

and where $R_{\min}$ is the value of the radius, R, that minimizes V^c at constant x_{OH} and salt concentration c_s.

Fig. 6 shows $V^c(R_{\min}, x_{\text{OH}}, c_s)$ as a function of x_{OH} at different values of the total salt concentration, c_s. The intersection of the curves with the line

$$y = (1 - x_{\text{OH}}) \ln x^-_2 - Q^c x_{\text{OH}} \qquad (44)$$

yields the points where $\beta = 0$, which is also in practice the condition to be met for long rod-shaped micelles to exist (to be more precise $\beta \lesssim 0.01$). However, since long rod-shaped micelles are much larger systems than spherical micelles they are practically all very nearly in equilibrium with the monomers in solution and we can write

$$kT \partial [(N_2 + N_3)\beta]/\partial N_2 = (\mu^-_{2,\text{mic}} - \mu^-_2)/kT \simeq 0 \quad (45)$$

and

$$kT \partial [(N_2 + N_3)\beta]/\partial N_3 = (\mu_{3,\text{mic}} - \mu_3)/kT \simeq 0 \quad (46)$$

or

$$\partial [(N_2 + N_3) V^c]/\partial N_2 = \ln x^-_2 \qquad (47)$$

and

$$\partial [(N_2 + N_3) V^c]/\partial N_3 = \ln x_3 - \varepsilon^c_{\text{OH}} \qquad (48)$$

i.e.

$$\ln x^-_2 = V^c(x_{\text{OH}}) - x_{\text{OH}}(\partial V^c/\partial x_{\text{OH}}) \qquad (49)$$

$$\ln x_3 - \varepsilon^c_{\text{OH}} = V^c(x_{\text{OH}}) + (1 - x_{\text{OH}})(\partial V^c/\partial x_{\text{OH}}) . \quad (50)$$

Thus, of all the lines described by Eq. (44) only the tangent line is relevant (Fig. 6).

Figure 6 shows that for some particular monomer concentration, x^-_2, by adding alcohol, rod-shaped micelles can be generated at salt concentrations far below those where rod-shaped micelles first appear in a pure surfactant solution. This is in full agreement, of course, with experimental findings [1] and is, by the way, hardly unexpected when, roughly speaking, a replacement of sulphate groups with OH groups is taking place on the micelle surface upon adding DOH. Such a replacement would obviously result in a lowering of the electrostatic free energy and, in addition, in the advantageous formation of a mixture. However, the free energy price to pay is largely due to the circumstance that DOH is present at a much lower monomer concentration than the DS^--ions.

There is an upper limit to x_3 set by the solubility of alcohol as a monomer in the solution and also an approximate upper limit to x_2^- somewhat above the CMC condition for the formation of spherical micelles at the actual salt concentration and the actual value of x_{OH}^{tot}. The values of c_s and x_{OH}^{tot} also determine the value of x_3 at the CMC.

The radius of the rod-shaped micelles does not vary much with the micelle mole fraction, x_{OH}, as shown in Table 2. Similarly to what was found in the case of spherical micelles a slight increase of R is noted when increasing the DOH content.

It was shown above that for small spherical micelles quite large fluctuations of the number of alcohol monomers will occur. However, at certain monomer concentrations (x_2^- and x_3) large rod-shaped micelles can actually exist for one single (rather well-defined) value only, of x_{OH}. This may be regarded as a consequence of the general relation between the size of the system and the magnitude of the fluctuations.

Table 2. Values of R_{min} [Å] for cylindrical SDS/DOH micelles of different composition

c_s (M)	$x_{OH} = 0$	$x_{OH} = 0.5$	$x_{OH} = 0.9$
0.005	14.45	14.87	15.15
0.01	14.50	14.90	15.16
0.05	14.63	14.98	15.18
0.1	14.69	15.01	15.19
0.5	14.88	15.08	15.19
1.0	14.89	15.11	15.19

Table 3. Effect of adding hexanol on the CMC of SDS as calculated in a more approximate manner assuming $\varepsilon_{OH}^s = 0$

c_2^- at the CMC (mM)	x_{OH}^{tot} (%)	x_3
8.348	0	0
7.500	19.96	0.3539×10^{-4}
6.500	41.10	0.8630×10^{-4}

These x_3 values are all well below the solubility of the hexanol in water ($x_3 = 1.0 \cdot 10^{-3}$).

Extension of the theory to alcohols with shorter chains

It seems that the above theory can be extended with reasonable accuracy to the case of shorter-chained alcohols with $n_c < 12$ if a number of minor adjustments are being made:

i)   The Tanford term $-(19.960 + \ln x_3)N_3$ should be replaced by the expression [17] $-(2.056 + 1.492 n_c + \ln x_3)N_3$.

ii) In the expression for the hydrophobic radius R and the conformational free energy, ε_{conf}, $N_{tot} = N_2 + N_3$ has to be replaced by $N_2 + n_c N_3/12$. This is, of course, a further approximation but owing to the dominating role of the free energy of mixing, this approximation should be satisfactory insofar as N_3 is sufficiently small.

iii) As a first approximation the free energy of mixing may be assumed to be

$$\Delta G_{mix} = N_2 \ln \left(\frac{12 N_2}{12 N_2 + n_c N_3} \right) + N_3 \ln \left(\frac{n_c N_3}{12 N_2 + n_c N_3} \right). \tag{51}$$

This expression is based on the volume fractions of surfactant and alcohol in a micelle. However, it is difficult to assess the degree of validity of this expression in the present case.

Calculations along these lines were made for spherical micelles containing hexanol ($n_c = 6$). Setting $\varepsilon_{OH}^s = 0$ the values shown in Table 3 for the CMC defined as the value of c_2^- where $\Phi_{tot} = 3 \times 10^{-4}$ were obtained. The x_3 values are all well below the solubility of the alcohol ($x_3 = 1.0 \times 10^{-3}$). Recently, similar calculations have been presented by Rao and Ruckenstein [4] where, however, the chain conformational free energy, ε_{conf}, and the electrostatic free energy, ε_{el}, were computed in less satisfactory ways.

Discussion

The treatment of mixed SDS/DOH micelles presented in this paper is founded on our previous theories for the formation of spherical and rod-shaped surfactant micelles [9, 10]. Our calculations here show that the addition of a long-chain aliphatic alcohol has a stabilizing effect on the micelles, as is obviously in line with well-established experimental facts. The origin of this stabilizing influence is rather complex.

Referring to the equilibrium micelle and the overall expression for $\varepsilon^s(N_2, N_3)$, Eq. (11), we may note that there is a free energy gain associated with the electrostatic term $N_2\varepsilon^s_{el}$ when N_2 is diminished in accordance with Eq. (30). The mixing term $\ln\begin{pmatrix} N_2+N_3 \\ N_3 \end{pmatrix}$ is, of course, always advantageous from the free energy point of view, i.e. it is <0. The remaining terms all yield positive contributions to $\varepsilon^s(N_2, N_3)$ when increasing N_3 and decreasing N_2. Thus, ε^s_{Tanf} becomes less negative because the monomer concentration of alcohol is generally less than that of the surfactant ion and, moreover, it is disadvantageous to add alcohol to the micelle surface as regards the head group term $N_2\varepsilon^s_{pg}+N_3\varepsilon^s_{OH}$. Finally, the chain configurational free energy term $(N_2+N_3)\varepsilon^s_{conf}$ and the "surface tension" term, $\gamma_{hc/w}aN_2/kT$, also increase somewhat when increasing the size of the hydrocarbon core.

In the case of spherical micelles, a straightforward comparison with experimental CMC data is hampered by the lack of information about how the CMC is actually defined by the experimental method employed. Still, a reasonable agreement has been demonstrated as to the lowering effect of DOH on the CMC of SDS upon assuming that there is no net free energy change associated with the transfer of an alcohol group from the monomer solution to the micelle surface ($\varepsilon^s_{OH} = 0$).

For rod-shaped micelles, the general picture is broadly the same. However, in this case the sensitivity to the detailed solution conditions is much more pronounced than for spherical micelles. Hence, by adding a long-chain aliphatic alcohol, the critical condition ($\beta = 0$) for the formation of rod-shaped micelles is approached at significantly lower salt concentrations.

It is noteworthy that when carrying out these calculations we have experienced no particular difficulties which can be attributed to the use of the constant $\gamma_{hc/w} = 50 \text{ mJm}^{-2}$ to model the contact free energy between hydrocarbon and water at the micelle/water interface. On this point it should be noted that we are in agreement with Rao and Ruckenstein [4] but not with Jönsson and Wennerström [3] who have used the value $\gamma_{hc/w} = 18 \text{ mJm}^{-2}$. This discrepancy will be further investigated in a forthcoming paper by the present authors. Finally, it appears from the way the calculations have been described above that experimental information on the monomer concentrations of surfactant, x_2^-, and of alcohol, x_3, and of the volume fraction of micelles, ϕ_{tot}, for mixed surfactant/long-chain alcohol micellar solutions would be of a considerable interest.

Acknowledgements

The authors wish to express their gratitude to Torbjörn Wärnheim for proposing the theme of this investigation and to Sten Sarman for pointing out that the exact combinatorial expression should be used when computing the entropy of mixing within the spherical micelles.

References

1. Ljosland E, Blokhus AM, Veggeland K, Backlund S, Høiland H (1985) Progr Colloid Polym Sci 70:34
2. Ekwall P (1975) Adv Liquid Crystals 1:1
3. Jönsson B (1981) The Thermodynamics of Ionic Amphiphile Water Systems. A Theoretical Analysis, Thesis, Lund; Jönsson B, Wennerström H, J Phys Chem. Submitted
4. Rao IV, Ruckenstein E (1986) J Colloid Interface Sci 113:375
5. Cabane B, Duplessix R, Zemb T (1985) J Physique 46:2161
6. Aniansson EAG, Wall SN, Almgren M, Hoffmann H, Kielmann I, Ulbricht W, Zana R, Lang J, Tondre C (1976) J Phys Chem 80:905
7. Hill, TL (1963 and 1964) Thermodynamics of Small Systems. Benjamin, New York, Vol. II
8. Israelachvili JN, Mitchell JJ, Ninham BW (1976) J Chem Soc, Faraday Trans 2, 72:1525
9. Eriksson JC, Ljunggren S, Henriksson U (1985) J Chem Soc, Faraday Trans 2, 81:833
10. Eriksson JC, Ljunggren S (1985) J Chem Soc, Faraday Trans 2, 81: 1209
11. Ljunggren S, Eriksson JC (1986) J Chem Soc, Faraday Trans 2, 82:913
12. Gruen DWR, DeLacey EHB (1984) Surfactants in Solution, vol 1, p 279, Mittal KL, Lindman B (eds). Plenum Press, New York
13. Gruen DWR (1985) J Phys Chem 89:153
14. Tanford C (1980) The Hydrophobic Effect, 2nd edn. Wiley, New York
15. Evans DF, Ninham BW (1983) J Phys Chem 87:5025
16. Miura M, Arichi S (1958) J Sci Hiroshima Univ Ser A, 22:57
17. Ref. [14], p 7

Received September 30, 1986;
accepted October 27, 1986

Authors' address:

Stig Ljunggren
Department of Physical Chemistry
The Royal Institute of Technology
S-10044 Stockholm, Sweden

Progress in Colloid & Polymer Science Progr Colloid & Polymer Sci 74:48–54 (1987)

Experimental evidence for repulsive and attractive forces not accounted for by conventional DLVO theory

P. M. Claesson

The Royal Institute of Technology, and The Institute for Surface Chemistry, Stockholm, Sweden

Abstract: Results from direct force measurements between mica surfaces and modified mica surfaces immersed in aqueous electrolyte solutions are reported. In particular, experimentally observed attractive and repulsive deviations from force curves, calculated from conventional DLVO theory, are discussed. Such "additional" forces are present in most systems. Even in cases where the measured interaction appears to be well described by conventional DLVO theory, the values of, for example, surface charge densities obtained by fitting calculated forces to measurements should not be viewed as true values.

Key words: DLVO theory, double-layer force, surface interaction, hydrophobic force, hydration force

Introduction

The purpose of this paper is to give some examples of how and when conventional DLVO theory fails to account for directly measured interaction forces between surfaces in aqueous electrolyte solutions. Conventional DLVO theory means double-layer forces in the non-linear Poisson Boltzmann (PB) approximation and van der Waals forces in the Lifshitz approximation.

Both repulsive and attractive forces not accounted for by the conventional DLVO theory are reported. "Additional" repulsive forces are observed between hydrophilic surfaces while "additional" attractive forces have been measured between hydrophobic surfaces, as well as between one hydrophobic positively charged surface and one negatively charged mica surface.

Some remarks concerning the common method of analysing force measurements are given in the light of recent theoretical calculations of double-layer forces [1–4], which have demonstrated severe shortcomings in the PB theory. A consequence of this is that surface charge densities and surface potentials obtained by fitting conventional DLVO forces to measurements are apparent and not true values [4].

Materials and methods

Chemicals

Muscovite mica sheets were obtained from Sciama, Paris. Dimethyldioctadecylammonium bromide was purchased from Eastman Kodak and recrystallized from ethanol. The KBr and NaI salts were from Merck and of suprapure grade. They were used without any further purification.

The water purification procedure included the following consecutive steps: decalcination, prefiltration with activated charcoal, followed by a treatment with a reverse osmosis unit, two mixed-bed ion exchangers, activated charcoal, Organex and a final filtration. All purification units were Millipore products, except for the final filter which was a Nucleopore 0.05 μm or a Zetapoore 0.2 μm filter. Before the water was used in a surface force experiment, it was deaerated for several hours.

Preparation of hydrophobic surfaces

Mica surfaces, coated with dimethyldioctadecylammonium ions (DDOA), were prepared using a standard Langmuir-Blodgett procedure. During deposition the area per molecule at the air/water interface was about 0.6 nm^2. After deposition, the area per DDOA molecule on the mica surface was determined by means of ESCA and found to be about 0.5 nm^2 [5]. The advancing water contact angle on such a DDOA-coated mica surface is 94°.

Measurements of surface interactions

Repulsive forces and the gradient of attractive forces were measured using the surface force apparatus of Israelachvili. These two force-measuring techniques are described in detail in references [6, 7].

The surface separation distance, D, between two molecularly smooth surfaces mounted in a crossed cylinder configuration was determined with an accuracy of about 0.1 nm using an interferometric technique [6]. The interaction force was measured from deflections of a double cantilever spring supporting the lower surface. The detection limit was about 10^{-7} N. The force between two crossed cylinders (F_c) is related to the free energy of interaction between flat surfaces, G_f [8, 9]

$$\frac{F_c}{R} = 2\pi G_f \tag{1}$$

where R is the local geometric mean radius.

The corresponding relation between the gradient of the force between crossed cylindrical surfaces and the pressure between flat surfaces, P_f, is:

$$\frac{1}{R}\frac{dF_c}{dD} = -2\pi P_f \tag{2}$$

Calculation of conventional DLVO forces

Double-layer forces were calculated from solutions to the exact Poisson-Boltzmann equation.

For similar surfaces, a numerical method developed by Chan et al. [10] was used. Calculated double-layer forces for dissimilar surfaces interacting at constant potential were obtained by using the elliptic integral formalism of Devereux and de Bruyn [11]. The corresponding double-layer forces for surfaces interacting at constant charge utilized the elliptic integral formulas of Bell and Peterson [12].

The non-retarded van der Waals force was calculated from:

$$\frac{F_c}{R} = \frac{A}{6D^2} \tag{3}$$

where the Hamaker constant A for mica interacting across water has been experimentally determined, by Israelachvili, to be $2\cdot2\cdot10^{-20}$ J [6]. Theoretically, the hydrocarbon layer present on DDOA-coated mica surfaces should reduce the van der Waals force at small surface separations.

Results and discussions

Figure 1 ilustrates the forces between two bare mica sheets measured in a range of aqueous NaI solutions. The conventional DLVO theory is commonly the starting point when analyzing such measurements. Among other things, the slope of the force curve is predicted by that theory, and at large distances given by the Debley length (K^{-1})

$$\kappa^{-1} = \left(\frac{\varepsilon_0 \varepsilon_r kT}{e^2 \sum_i n_i(B) Z_i^2} \right)^{1/2} \tag{4}$$

where $n_i(B)$ is the number density of ion i in bulk solution. Debye-lengths determined from force measurements in simple electrolytes generally agree well

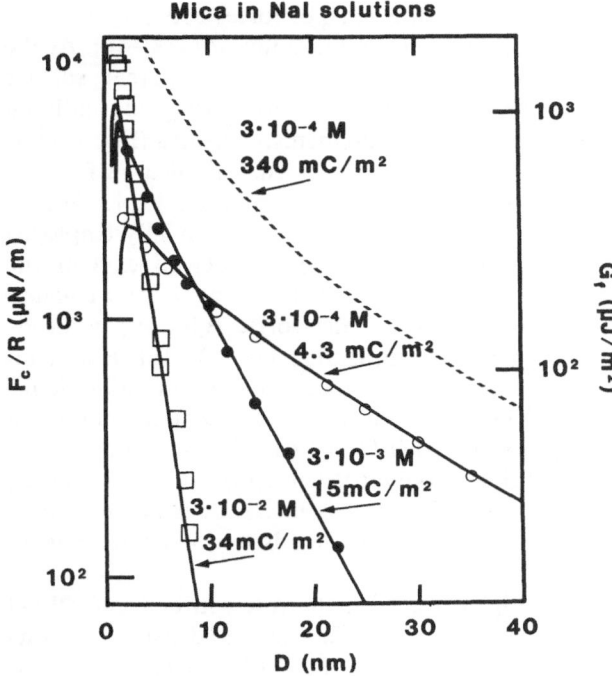

Fig. 1. The forces measured between two crossed cylindrical bare mica surfaces immersed in a range of NaI solutions as a function of surface separation. (○) represents measured forces in $3\cdot10^{-4}$ M solutions, (●) forces in $3\cdot10^{-3}$ M and (□) forces in $3\cdot10^{-2}$ M. (—) represent conventional DLVO forces for surfaces interacting at constant charge which are best fits to measurements. (− − −) represents the expected DLVO forces if no adsorption of counterions onto the mica surfaces occurred

with predictions based on the conventional DLVO theory [13–16]. This is also the case for the force curves presented in Fig. 1.

The magnitude of the double-layer force at a given electrolyte concentration depends on the surface charge densities. The dashed line in Fig. 1 represents predictions of the conventional DLVO theory for two surfaces, each having a surface charge density of -0.34 C/m² interacting in a $3\cdot10^{-4}$ M 1:1 electrolyte solution. This surface charge density corresponds to the muscovite mica lattice charge originating from the circumstance that $2.1\cdot10^{18}$ silicon atoms/m² are replaced by aluminum atoms [17].

As is illustrated in Fig. 1, the magnitude of the forces measured between mica surfaces in a $3\cdot10^{-4}$ M NaI solution is considerably less. The apparent surface charge density, obtained by fitting conventional DLVO theory for surfaces interacting at constant charge to the measured forces, is -4.3 mC/m². The small magnitude of the surface charge density, compared with the lattice charge, is in qualitative accordance with earlier measurements of forces between mica surfaces in different electrolyte solutions [13−16]. As the NaI concentration increases, the apparent surface charge density, determined by fitting conventional DLVO forces to measurements, increases (Fig. 1). The same trend has been observed for mica surfaces in other monovalent electrolyte solutions [6, 13, 16].

Pashley and coworkers have successfully employed an ion adsorption exchange model to explain the variation of the mica surface charge density (surface potential) in different electrolyte solutions as determined from force measurements [13−16]. ESCA investigations of mica surfaces after equilibration in different electrolyte solutions, show that the proposed adsorption model in many cases reasonably predicts ion densities on the muscovite mica surface. However, it is also clear that such a model does not provide a complete picture of the ion adsorption characteristics [18].

From Fig. 1 it is clear that the conventional DLVO theory, with a fitted surface charge density, accounts for the measured forces in $3 \cdot 10^{-4}$ M and $3 \cdot 10^{-3}$ M NaI solutions. In particular, the van der Waals force overcomes the repulsive double-layer force 2−3 nm from bare mica-mica contact. The adhesion force at $D = 0$, normalized with the local geometric mean radius of the interacting surfaces, is about 50 mN/m. However, in $3 \cdot 10^{-2}$ M NaI solutions a strong repulsive force prevents the surfaces from reaching molecular contact. This is not predicted by conventional DLVO theory. Consequently, this short-range $(D < 5 \text{ nm})$ repulsive force is commonly viewed as an additional repulsive force. Such forces have been named hydration forces. They have been observed between mica surfaces in a range of electrolyte solutions [13−16], mica surfaces coated with surfactant and lipid bilayers [19−21], and between lipid multibilayers [22]. A similar solvation force has also been measured between lipid multibilayers in ethyleneglycol [23]. The molecular origin of the repulsive hydration force is not clear. However, it has been suggested that a structural force due to dehydration of polar groups [13] or a force due to polarization charges [24] might explain the measured hydration force.

At this stage, one has to consider how accurately conventional DLVO theory predicts double-layer and van der Waals forces. Recent calculations and simulations of double-layer forces have shown that predictions based upon the non-linear Poisson-Boltzmann (PB) equation are quantitatively or even, in some cases, qualitatively wrong [1−2].

Ion-ion correlations, ionic size effects and polarization charges all contribute to real double-layer forces

but these contributions are neglected in the PB model. For instance, Kjellander and Marčelja have shown, by solving the anisotropic hypernetted chain equation, that double-layer forces between surfaces in monovalent electrolyte solutions of small ions are less repulsive than predicted by the PB theory [4]. This is due to ion-ion correlations. However, they also found that the shape of the force curve is, in many cases, hardly affected by ion-ion correlations. Hence, in such cases it should be possible to fit real double-layer forces with PB theory by assuming a too low surface charge density [4].

Ion-ion correlations are hardly the only reason for the difference between the forces actually measured between mica surfaces in dilute electrolyte solutions and those predicted by conventional DLVO theory, assuming no adsorption of counterions (Fig. 1). Instead, it appears that counterions are adsorbed onto the muscovite mica surface [18], but nevertheless it is clear that the resulting surface charge densities cannot be obtained by fitting conventional DLVO theory to measured forces. Surface charge densities obtained by such a fitting procedure should instead be viewed as apparent values. Consequently, adsorption models proposed to explain such apparent values [13−16, 18] are not expected to describe the reality completely.

The common way of calculating additional non-DLVO forces is to subtract forces calculated by conventional DLVO theory from the total measured force. This method is somewhat artificial. One reason for this is that real DLVO forces are not accurately predicted by conventional DLVO theory, as discussed above. Another reason is that only the total force is measurable and constitutes a well defined concept. Any division into different force contributions is an interpretation based on theoretical calculations of the strength and distance dependence of such force contributions. Such a division is only an approximation, since different types of forces are not independent. For instance, ion-ion correlations within the double layer give rise to a van der Waals type of interaction [25, 26]. Hence, the distinction between double-layer forces and van der Waals forces is somewhat artificial. Furthermore, the presence of a structural force implies that the solution adjacent to a surface has a somewhat different density and/or dielectric properties than the bulk solution. This will, of course, also affect double-layer forces and van der Waals forces [9]. Once again, one has to conclude that the distinction between different force contributions is somewhat artificial. Nevertheless, such a division is often a good approximation and it provides the means for distinguishing and discussing different aspects of the total force.

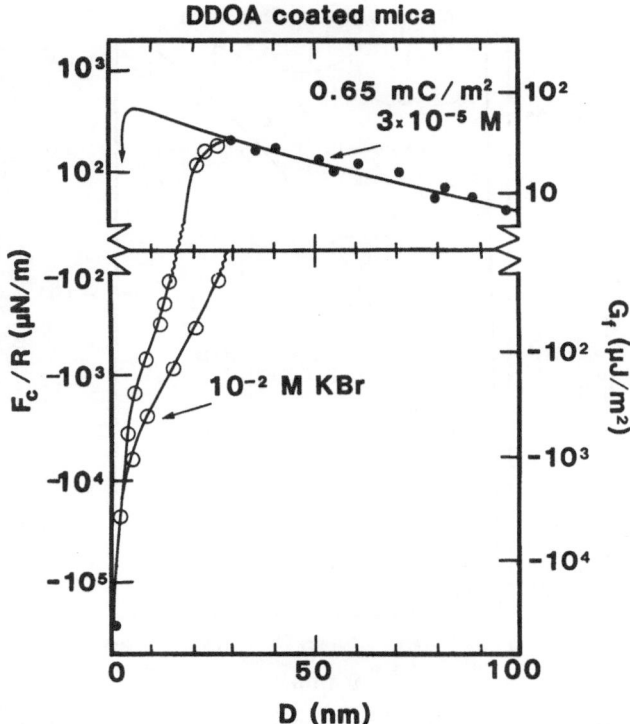

Fig. 2. The force between two DDOA-coated mica surfaces as a function of surface separation. (●) represents directly measured forces and (○ forces obtained by integrating the directly measured slope of the force. The upper solid line represents calculated conventional DLVO forces for surfaces interacting at constant charge ($\sigma_0 = 6.5 \cdot 10^{-4} \, C/m^2$, $\psi_0^\infty = 45 \, mV$). The Van der Waals force has been assumed to be the same as for uncoated mica surfaces

Hence, the force curves reported here are discussed in terms of different force contributions, although it is realized that this is an approximation.

Figure 2 illustrates the forces between two mica surfaces which have been rendered hydrophobic by deposition of a Langmuir-Blodgett monolayer of dimethyl-dioctadecylammonium ions (DDOA). The deposited monolayer is 2.0 ± 0.2 nm thick and stable in aqueous solutions [25]. The area per deposited DDOA is about $0.5 \, nm^2$, as determined by ESCA.

In dilute electrolyte solutions the measured forces at large separations ($D \geq 30$ nm) are, at least apparently, well described by conventional DLVO theory. However, at smaller distances, conventional DLVO theory (upper solid line) completely fails to predict the measured interaction (middle solid line). Instead of the expected repulsive force, a strong attraction is observed.

In 10^{-2} M KBr a purely attractive force is observed between DDOA-coated mica surfaces, while conven-

tional DLVO theory predicts the force curve to be repulsive over a range of distances. As is illustrated in Fig. 2, the total force clearly changes with the addition of salt. However, the additional force obtained by subtracting the force calculated from DLVO theory from the total force shows only a weak salt dependence [25]. This additional attractive force, which is close to two orders of magnitude stronger than the expected Lifshitz-van der Waals force, is the hydrophobic interaction between surfaces. Its molecular origin is not known, but a force due to the overlap of regions with a surface-enhanced water structure has been suggested [25–26]. In this context, it should be noted that two hydrophobic surfaces ($\theta > 90°$) separated by a thin water film constitutes a thermodynamically metastable state [27–28]. In the stable state, a cavity filled with water vapour should exist between the hydrophobic surfaces. However, the energy barrier for cavity formation is very large whenever the surfaces are separated by more than a few tenths of a nanometer [28].

The formation of a cavity between the surfaces would result in a change in the refractive index of the separating medium. Such a change would show up as a discontinuity in the standing waves, fringes of equal chromatic order, used for determining the surface separation distance interferometrically. No such discontinuities are observed as the surfaces approach each other which is when the interaction, shown in Fig. 2, is measured.

However, recent measurements of the interaction between mica surfaces coated with a double-chained fluorocarbon-based surfactant ($2C_8F_{17}C_2H_2$-L-glu-$CH_2-N-(CH_2)_3$ Br, Sogo pharmaceutical Co Ltd), carried out in this laboratory have shown that a cavity does form between such surfaces when in molecular contact [29]. Preliminary measurements of the interaction between two such surfaces, measured when approaching each other with no cavity present, indicate an attraction similar to the one observed between DDOA-coated mica surfaces (Fig. 2). Once the cavity has formed (with the surfaces in molecular contact) it is possible to separate the surfaces several micrometers before the cavity disappears. A more detailed discussion of this phenomenon will be reported elsewhere. Anyway, the hydrophobic interaction between surfaces as determined by direct force measurements appears to be a phenomenon related to, but not necessarily caused by, cavity formation.

Figures 3–5 illustrate the interaction measured between one negatively charged mica surface and one positively charged hydrophobic DDOA-coated mica surface. The attractive interaction in the mixed system

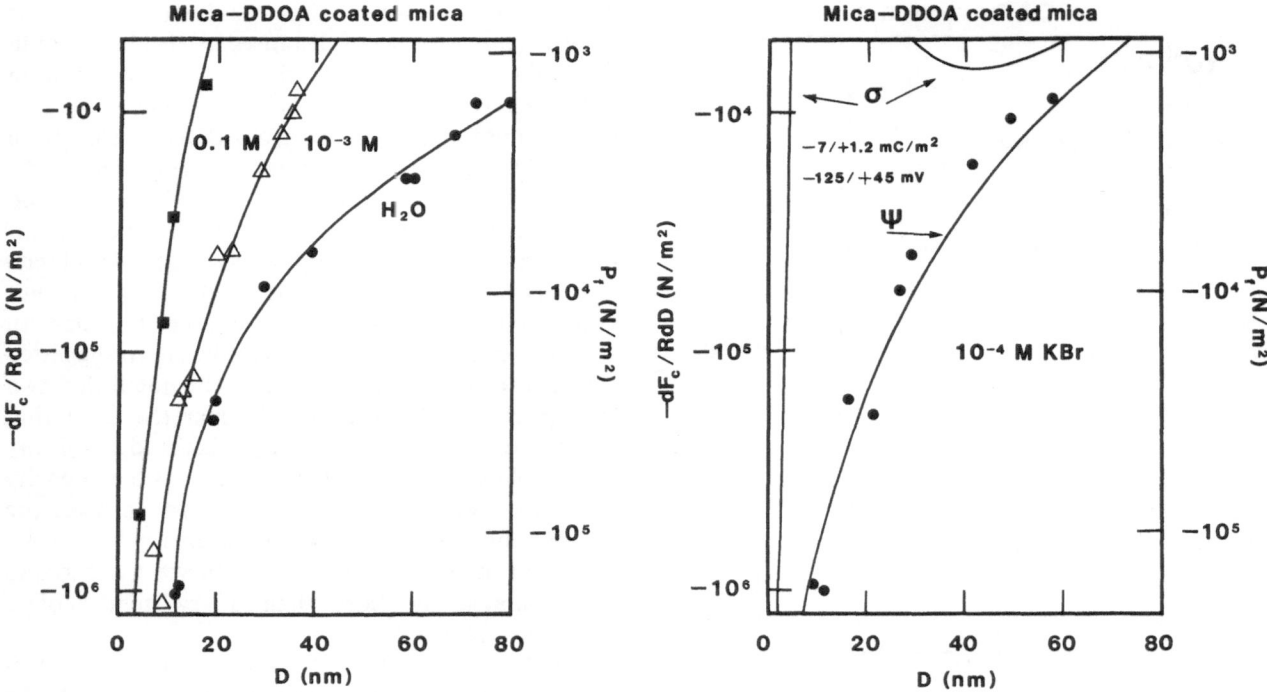

Fig. 3. The slope of the force between one negatively charged bare mica surface and one positively charged hydrophobic DDOA-coated surface, immersed in a range of KBr solutions, as a function of surface separation.
(■) represents measurements in 0.1 M solutions, (△) measurements in 10^{-3} M solutions and (●) measurements in pure water. The solid lines are guidelines only

Fig. 4. The slope of the force between one bare negatively charged mica surface and one positively charged DDOA-coated mica surface, as a function of surface separation. Measurements are represented by (●). The upper and lower solid lines represent force curves, calculated from conventional DLVO theory, for surfaces interacting at constant charge and constant potential, respectively

Table 1. Apparent surface potentials and surface charge densities.

C (M)	ψ_0(mV)	σ_0(C/m²)	A/charge
10^{-4}	− 125	− 0.0070	2278
	+ 45	0.0012	13254
10^{-3}	− 125	− 0.022	721
	+ 45	0.0038	4197
$3 \cdot 10^{-3}$	− 100	− 0.023	691
	+ 45	0.0066	2420
10^{-2}	− 60	− 0.018	903
	+ 30	0.0075	2134
0.1	− 125	− 0.22	72
	+ 55	0.050	322

The surface charge densities and surface potentials obtained by fitting conventional DLVO theory to measurements (in 0.1 M KBr no fit to experiment was possible). As discussed in the text, these values might not correspond to the real values because of limitations of the DLVO theory

is strongly salt dependent (Fig. 3), which contrasts with the weak salt dependence of the hydrophobic interaction between two DDOA-coated mica surfaces.

Forces measured in dilute KBr solutions ($C \leq 10^{-3}$ M) can be fitted by conventional DLVO theory, assuming interaction at constant potential, as is illustrated in Fig. 4. It is puzzling that the interaction appears to occur under constant potential conditions, since it implies that the surface charge densities increase with decreasing surface separation [30].

Two related difficulties which also arise when trying to fit conventional DLVO theory to these measurements are illustrated in Table 1 and Fig. 5. Table 1 shows that in order to fit conventional DLVO theory to the measured attraction, larger surface charge densities have to be assumed at higher electrolyte concentrations. It is hard to find an explanation for an increased surface charge density on the DDOA-coated surface, prepared by an LB technique. This is discuss-

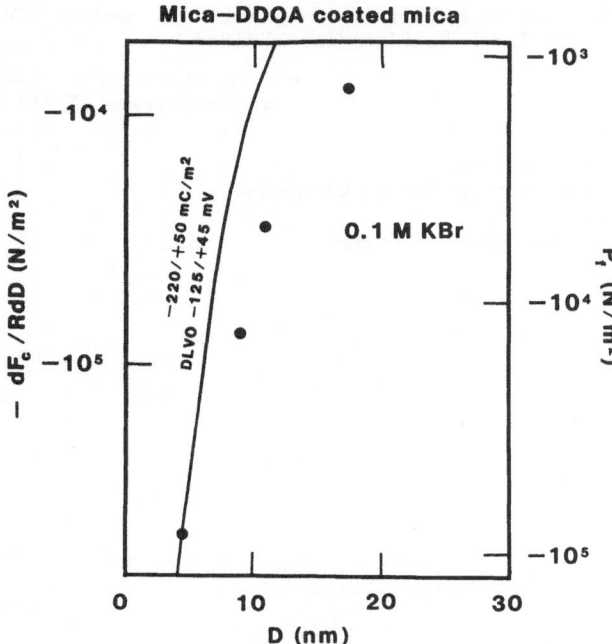

Fig. 5. The slope of the force between one negatively charged bare mica surface and one positively charged DDOA-coated mica surface as a function of surface separation. (●) represents measurements in 0.1 M KBr solutions. The solid line represents a force curve calculated from conventional DLVO theory. Clearly, the expected attractive DLVO force (even with assumptions of unreasonable, large surface charge densities) is of a too short-range to account for the measured attraction

from complete theory for particle and surface interactions. Repulsive deviations from predictions based on conventional DLVO theory are observed for interacting hydrophylic surfaces, while attractive deviations are measured between hydrophobic surfaces. The shortcomings of the conventional DLVO theory are also illustrated by considering the interaction between dissimilar surfaces, for instance, one bare mica surface and one DDOA-coated mica surface. Furthermore, the apparent success of the conventional DLVO theory in explaining some force measurements between similar surfaces might, to a large extent, be coincidental.

ed in ref. [31]. Furthermore, the attraction measured in 0.1 M KBr solutions are too long-range to be caused by an attractive double-layer force, as is illustrated in Fig. 5.

Hence, one can conclude that conventional DLVO theory fails to consistently explain the measured interaction between one bare negatively charged mica surface and one hydrophobic DDOA-coated mica surface with a small positive charge. Instead, the strong attraction observed must be due to other force contributions as well. Such contributions might arise from ion-ion correlations and/or from an interaction related to the hydrophobic interaction observed between two DDOA-coated surfaces [25].

Conclusions

Conventional DLVO theory i.e. double-layer forces in the Poisson-Boltzmann approximation and van der Waals forces in the Lifshitz approximation, is a far

References

1. Guldbrand L, Jönsson B, Wennerström H, Linse P (1984) J Chem Phys 80:2221
2. Kjellander R, Marčelja S (1984) Chem Phys Lett 112:49
3. Kjellander R, Marčelja S (1985) J Chem Phys. 82:2122
4. Kjellander R, Marčelja, S (submitted) J Chem Phys
5. Herder PC, Claesson PM, Blom CE (in press) J Colloid Interface Sci
6. Israelachvili JN, Adams GE (1978) J Chem Soc Faraday Trans I 74:975
7. Israelachvili JN, Pashley RM (1984) J Colloid Interface Sci 98:500
8. Derjaguin BV (1934) Kolloid 69:155
9. Israelachvili JN (1985) "Intermolecular and Surface Forces with Applications to Colloid and Biological Systems", Academic Press, London
10. Chan DYC, Pashley RM, White LR (1980) J Colloid Interface Sci 77:283
11. Devereux OF, de Bruyn PL (1963) "Interaction of Plane-Parallel Double Layers", The M.I.T. Press, Cambridge, Massachusetts
12. Bell GM, Peterson GJ (1972) J Colloid Interface Sci 41:542
13. Pashley RM (1981) J Colloid Interface Sci 83:531
14. Pashley RM, Israelachvili JN (1984) J Colloid Interface Sci 97:446
15. Pashley RM (1984) J Colloid Interface Sci 102:23
16. Claesson PM, Horn RG, Pashley RM (1984) J Colloid Interface Sci 100:250
17. Güven N (1971) Zeitschrift Krystallographic 134:196
18. Claesson PM, Herder PC, Stenius P, Eriksson JC, Pashley RM (1986) J Colloid Interface Sci 109:31
19. Pashley RM, Israelachvili JN (1981) Colloids and Surfaces 2:169
20. Claesson PM, Kjellander RK, Stenius P, Christenson HK (1986) J Chem Soc Faraday Trans I 82:2735
21. Marra J, Israelachvili JN (1985) Biochemistry 24:4608
22. Parsegian VA, Fuller N, Rand RP (1979) Proc Natl Acad Sci USA 76:2750
23. Persson PKT, Bergenståhl BA (1985) Biophys J 47:743
24. Jönsson B, Wennerström H (1983) J Chem Soc Faraday Trans II 79:19
25. Claesson PM, Blom CE, Herder PC, Ninham B (1986) J Colloid Interface Sci 114:234

26. Israelachvili JN, Pashley RM (1984) J Colloid Interface Sci 98:500
27. Yaminsky VV, Yushchenko VS, Amelina EA, Shchukin ED (1983) J Colloid Interface Sci 96:307
28. Yushchenko VS, Yaminsky VV, Shchukin ED (1983) J Colloid Interface Sci 96:307
29. Christenson HK, Claesson PM, Herder PC, Berg J (in preparation)
30. Chan D, Healy TW, White LR (1976) J Chem Soc Faraday Trans I 72:2844
31. Claesson PM, Herder PC, Blom CE, Ninham BW (1987) J Colloid Interface Sci 118:68

Received September 8, 1986;
accepted October 27, 1986

Author's address:

P. M. Claesson
The Institute for Surface Chemistry
P.O. Box 5607
S-11 486 Stockholm, Sweden

Progress in Colloid & Polymer Science   Progr. Colloid & Polymer Sci 74:55–63 (1987)

Migration of small hydrophobic molecules between micelles in aqueous solution

M. Almgren and J. Alsins

The Institute of Physical Chemistry, University of Uppsala, Uppsala, Sweden

Abstract: Triplet energy transfer from 9-methylanthracene to azulene or guaj-azulene has been used to probe the migration of azulenes between micelles in aqueous solution. The migration of the hydrophobic solutes between small ionic and nonionic micelles had the temperature dependence expected for a process controlled by diffusion through the intermicellar solution, although the rate in some cases was substantially less than calculated from the Smoluchowski equation. Under conditions in which the micelles grow into large, probably rod-like structures, there are severe difficulties in separating the inter- and intramicellar deactivation processes. The intermicellar migration was enhanced under these conditions, in cetyltrimethylammonium surfactants on addition of chlorate ions, and in hexaethylene glycol dodecylether at temperatures approaching the cloud-point. The mechanism of this migration is discussed and compared with pertinent results from micelle relaxation kinetics and surfactant self-diffusion measurements.

Key words: Triplet energy transfer, micelle, solubilization, kinetics

Introduction

Exchange of surfactant monomers or solubilized molecules between small ionic micelles seems to occur exclusively via the intermicellar solution [1]. Exchange mechanisms involving micelle collisions are excluded by the strong intermicellar repulsion. At high salt concentrations and for nonionic surfactants collisions between micelles should be much more frequent, and exchange on collisions or in micelle fusion-fission processes would be more probable. Indeed, Kahlweit et al. [2–4] showed that fusion-fission processes dominate the kinetics of micelle formation and breakdown at high concentrations of counterions. For nonionic surfactants Nilsson et al. [5] proposed that direct exchange of surfactant monomers in micelle collisions contributes to the self-diffusion process, under certain conditions.

Recently, Zana et al. [6, 7] claimed that a fast exchange of very hydrophobic molecules like pyrene could occur between nonionic micelles [7] and in ionic micellar solutions containing high concentrations of counterions or strongly bound counterions [6]. They measured the fluorescence decay of pyrene under conditions when quenching or excimer formation could occur. With immobile probes and quenchers, the decay should reach a final exponential phase with a decay time equal to that of unquenched pyrene; this fluorescence emanates from micelles containing only an excited pyrene molecule and no additional probe or quencher. The final part of the observed decay was faster, however, indicating a migration of the probes or quenchers. The residence time for pyrene in the micelles was estimated as 0.3 to 10 µs, much shorter than that found for pyrene in small ionic micelles — 100 to 1000 µs [1]. To explain the rapid migration, Zana et al. invoked exchange on micelle collisions or, in the case of ionic micelles, a transport of the hydrophobic solute in a small micelle fragment supposed to split off from one large micelle and merge with another. Zana et al. suggested that these processes would be important under the same conditions as the fusion-fission processes for the micelle kinetics, i.e. when the solution contains long rodlike micelles with a broad size distribution. Such processes would also contribute to the self-diffusion of the solutes and the surfactant molecules.

The migration processes discussed by Zana et al. are interesting and physically sound, but still hypothetical. All evidence for them stem from fluorescence decay data analyzed with the Infelta model [8]. Strictly, this model is valid only for quenching in a system of monodisperse micelles, and allows for migration of quenchers only. It is possible that the feature of the decay that suggested a migration was, in reality, due to intramicellar quenching in large, polydisperse micelles.

The migrative processes will deactivate all excited states in the micellar solution, and are final deactivation processes, therefore. Those suggested by Zana et al. are general; all probes and quenchers would exchange with similar rates by these mechanisms. It should be comparatively simple to falsify the hypothesis, therefore: it suffices to show that the final deactivation of a probe by deactivators in the micellar solution can occur much slower than the final deactivation of pyrene observed in the experiments of Zana et al. Such slow deactivation processes can be detected only if more long-lived excited probes are used. We have studied the deactivation of the triplet state of 9-methylanthracene, therefore, by triplet energy transfer to azulene. The lifetime of the triplet can reach a millisecond or more in carefully deareated micellar solutions, providing an appropriate time-window for the study of migration processes on the micro- to millisecond timescale.

Experimental

Materials

Sodium dodecylsulfate, SDS, BDH special purified grade, was used as supplied. Dodecyloxyethylene monoethers, $C_{12}E_6$ and $C_{12}E_8$, were high purity samples obtained from Nikko Chemical Co., Tokyo. Two samples of cetyltrimethylammonium chloride, CTAC, were used in the measurements. One was a gift from a donor who prefers to remain anonymous. On analysis, this preparation contained 85% Br^- as counterion and was used in the measurements with azulene. The other was prepared from CTAB (Merck, No. 2342, p.a.) by ion exchange on a Dowex 1-X8 resin, and recrystallized several times from ethylacetate-ethanol (9:1). This sample, used in measurements with guajazulene and in solubilization studies, also contained some Br^-, about 8%. The presence of Br^- does not affect the photophysics, but it has an effect on the size of the micelles, increasing the tendency of rod formation. 9-Methylanthracene (MeA) from Aldrich, was recrystallized twice from ethanol. Azulene (Az) and guajazulene (GAz) (1,4-dimethyl-7-isopropylazulene) were both from Aldrich and used without further purification. Deoxygenation of solutions was made by gentle bubbling with argon in a high stem cuvette.

Flash photolysis

An excimer laser, Lambda Physik EMG 100, provided a 20 ns flash at 351 nm (XeF). Analyzing light from a pulsed Xe lamp passed a 400 nm cut-off filter, Schott KV370 and GG400 (1 mm), and entered the 1 cm optical cell at right angles to the excitation direction, just behind the window facing the excitation laser. A Zeiss double monochromator, MM 12, with quartz prisms, was used for wavelength selection. In the measurements on MeA the intense triplet-triplet absorption around the maximum at 427 nm was followed. The detector was usually a Hamamatsu R928 photomultiplier with reduced number of active dynodes. To avoid ionization of MeA the laser beam was defocused until the ratio of the absorption signals at 700 nm (solvated electron) and 427 nm (MeA triplet) became constant.

The signals were captured with a Tektronix 7912 AD digitizer. Signal processing and averaging were performed on a Tektronix 4052 microprocessor, which was also used for analyzing the recorded decays by nonlinear curve-fitting to various models.

Solubilization measurements

The distribution of Az between micelles and the aqueous solution was estimated by determination of the solubilities at several temperatures, both in the micellar solutions and in the solvents (inclusive of added salt). Solid azulene was equilibrated with the solution in a 1 cm or 0.2 cm optical cell placed in an air thermostat and slowly rotated to provide stirring. Suspended azulene particles were removed by centrifugation inside the thermostat. The cell was then transferred to a thermostatted cell holder in a Zeiss DMR 10 absorption spectrometer. The procedure was tested for SDS by equilibration from temperatures both above and below the final equilibrium temperature, with results in very close agreement.

The absorption spectrum of azulene between 400 and 800 nm was similar in ethanol and the surfactant solutions, whereas the absorption spectrum in water was similar to that from water-ethanol 5:1. Molar absorptivities from ethanol was used for the surfactant solutions, and from water-ethanol for the water solutions.

Azulene slowly decomposes in aerated solutions, as evidenced by absorption changes which were largest at short wavelengths but still significant around 575 nm. No change was observed at 625 nm where the azulene concentration was determined.

Results and discussion

Intra- and inter-micellar deactivation

In solutions of small monodisperse micelles the deactivation of excited states occurs in two distinct stages [8], reflecting the rapid *intra*micellar deactivation in micelles containing quenchers followed by a slower quenching due to *inter*micellar migration. The triplet donor in the present study, 9-MeA, is very hydrophobic with an expected residence time in small micelles approaching 1 ms, as for anthracene [1]. The acceptor, azulene, similar to naphthalene in its water solubility, would migrate much faster by diffusion through the intermicellar aqueous solution. Since the triplet energy of MeA is higher than that of Az the energy transfer should occur on every encounter of the

MeA-triplet and Az. The intramicellar process should be very rapid, therefore, with a quenching constant k_q similar to that observed in diffusion-controlled fluorescence quenching in small micelles, which means a value around 10^7 s^{-1} or even larger, at room temperature. The time-resolution of the triplet absorption spectroscopy in our set-up was of the order of 10^{-7} s, which means that the rapid intramicellar deactivation was not well resolved for small micelles, and only the subsequent slow decay studied.

Migration mechanisms involving micelle collisions become more favorable when the conditions are changed so that the intermicellar repulsion is reduced. The repulsion between the headgroups on each micelle is then also reduced, favoring the growth of large, usually rod-like micelles, with pronounced polydispersity. The growth of the micelles makes the intramicellar process slower and the increased polydispersity complicates the interpretation of the experimental results.

Excited state decay in monodisperse micellar solutions is usually described by an equation, first derived by Infelta et al. [8] for fluorescence decay. This equation allows for migration of the quencher between the micelles. An equation of the same form but with a different interpretation of the parameters is a good approximation for the decay with any migration mechanism, involving excited probe, quencher, or both, in a monodisperse micellar solution [9]:

$$F(t) = A_1 \exp\{-A_2 t + A_3 [\exp(-A_4 t) - 1]\} \qquad (1)$$

where $\quad A_2 = k_0 + k_q \langle x \rangle_s$
$$A_3 = n(1 - \langle x \rangle_s / n)^2$$
$$A_4 = k_q (1 - \langle x \rangle_s / n)^{-1}$$

A_1 is the fluorescence intensity at time zero (the moment of excitation), k_0 the decay constant of the excited probe, n the mean number of quenchers per micelle, k_q the first order rate constant for quenching in a micelle with one quencher, and $\langle x \rangle_s$ the mean number of quenchers in micelles that still contain excited states during the final exponential decay. In the Infelta case [8]

$$\langle x \rangle_s = k_- n / (k_- + k_q)$$

where k_- is the exit rate constant for a quencher from a micelle.

Equation (1) should also apply to the decay of the triplet absorption signal in monodisperse micellar solutions. In our measurements, however, the intramicellar process was well-resolved only when the micelles were large and probably very polydisperse.

Equation (1) does not apply, therefore, and the results will be discussed in a more qualitative way. The initial rapid deactivation can be charaterized roughly by the initial decay parameter, k_i, defined by

$$k_i = -\lim_{t \to 0} (\text{d} \log F(t) / \text{d}t) \qquad (2)$$

which can be shown from Eq. (1) to be given by

$$k_i = A_2 + A_3 A_4 = k_0 + k_q n \ . \qquad (3)$$

When conditions such as salt concentration and temperature are changed so that the micelles grow at constant surfactant and quencher concentrations, n is expected to be proportional to the aggregation number, N_s. In the size regime where the micelles are globular k_q is expected to be inversely proportional to N_s, for both experimental [10–12] and theoretical reasons [13, 14]. The initial decay rate is then independent of the micelle size, and the main effect of the growth of the micelles in a decrease in amplitude of the slow decay. When the micelle size increases further into the regime of long rods n continues to grow in proportion to N_s, but k_q will now decrease faster. For long rods k_q will be inversely proportional to the square of the length, or proportional to N_s^{-2}. The initial decay rate constant will then decrease as N_s^{-1}.

The growth of the micelles into long rods will, therefore, be accompanied by a slowing down of the intramicellar quenching events and a decrease in the amplitude of the slowest decay process. The polydispersity is expected to increase simultaneously and the intramicellar processes may therefore spread out extensively in time, and prevent a clear separation between intra- and inter-micellar events. The use of very long-lived probes is essential under such circumstances.

The decay constant of the intermicellar deactivation process will be denoted k_s, and is equal to the term $k_q \langle x \rangle_s$ of parameter A_2 of Eq. (1). In the Infelta case

$$k_s = k_q k_- n / (k_q + k_-) \ .$$

This instructive expression shows that when the intramicellar quenching is slow compared to the exchange, k_s reduces to $k_q n$, i.e. the same as the initial decay constant. The whole decay is then exponential as in a homogeneous solution. If the quenching is fast compared to the exchange we get

$$k_s = k_- n = k_+ [Q]_{\text{free}} \qquad (4)$$

where k_+ is the entrance rate constant and $[Q]_{free}$ the concentration of quencher in the intermicellar solution. Equation (4) is expected to apply for 9-MeA, Az in small ionic micelles, which we will refer to as a normal case. Earlier studies [1] have shown that k_+ is close to the diffusion-controlled value, according to the Smoluchowski equation

$$k_{diff} = 4\Pi(R_m + R_q)(D_m + D_q)N_A \qquad (5)$$

where R is radius, D diffusion coefficient, and subscripts m and q signify micelle and quencher, respectively. For concentrated micelle solutions, where the distance between the micelles becomes comparable to R_m, some increase in k_{diff} beyond Eq. (5) would be expected. It is more difficult to assess a value for k_{diff} in a solution of rod-like micelles. It will increase above the value for spheres, but certainly not in proportion to the length. In any case, it is expected that the diffusion-controlled migration rate will increase somewhat when rods are formed, provided that the distribution of the reactant between the micelles and the aqueous subphase remains unchanged.

It is much more difficult to handle other exchange mechanisms. Interpreting experimental data with Eq. (1) gives a value of $\langle x \rangle_s$, from which parameter values of the pertinent rate-constants can be deduced, if the micelles are monodisperse and the exchange mechanism known [9]. Since neither condition is fullfilled we will resort to comparisons of observed k_s values under various conditions.

The rate of direct exchange on micelle collisions, or in micelle fusion-fission processes, should be rather insensitive to the nature of the exchanging species. Molecules of different hydrophobicity would exchange about equally fast; also the surfactant monomer itself. Information about the latter is available from two sources: studies of micelle relaxation kinetics and self-diffusion.

From their experimental results for the slow relaxation process in ionic micelles at high counterion concentrations, Kahlweit and coworkers [2–4] estimated a value of $20 \, s^{-1}$ for the mean dissociation rate constant of a micelle into two smaller ones. The slow relaxation process for nonionic surfactants seems to occur on a similar timescale [15]. These processes cannot be responsible for migration on the microsecond timescale. Micelle collisions can occur much more frequently, however, and may lead to an exchange of some monomers or hydrophobic solutes, without a complete merging of the individual micelles. Such collisions would not lead to a complete relaxation of the micelle equilibrium – the number of micelles is not changed – but would contribute to both migration and self-diffusion.

The self-diffusion data of interest in this connection were obtained by Nilsson et al. [5] for nonionic micelles of $C_{12}E_5$ and $C_{12}E_8$ in the L_1 phase. The self-diffusion coefficients of both surfactants decrease strongly with concentration initially, indicating an immobilization of the micelles due to strong interactions and/or growth in size. At higher concentrations, however, the self-diffusion was slightly enhanced for both surfactants, in a temperature region down to $25 \, °C$ below their cloud points. It is possible that a self-diffusion mechanism based on exchange of monomers is responsible for the break of the downward trend. To be a little more specific, consider a situation where the micelles are almost stationary, each confined to stay within a cage spanned by its neighbors. The micelle moves within its cage, colliding and occasionally exchanging monomers with the neighbors at a frequency k_{ex}. The contribution to the self-diffusion from this process would be

$$D_{ex} = L^2 k_{ex}/6 \qquad (6)$$

where L^2 is the average square distance between two neighbors. If the micelle size remains constant with increasing surfactant concentration, L^2 should be proportional to $[mic]^{-2/3}$. The exchange frequency is expected to be proportional to the micelle concentration and thus

$$D_{ex} \sim [mic]^{1/3}$$

which is close to the observed dependence.

Equation (6) can be used to compare self-diffusion and migration data. D_{ex} is one of several contributions to the measured self-diffusion coefficient, and k_{ex} is approximately related to the experimental decay parameters by k_s/n. For $C_{12}E_8$ micelles Zana and Weill [7] reported aggregation numbers and rate constants calculated from fluorescence quenching data. The first order rate constant for the intermicellar migration of pyrene corresponds to k_s and gives $k_{ex} = 2 \times 10^6 \, s^{-1}$ (at $66.5 \, °C$) at a surfactant concentration of approximately 3%, corresponding to a mean distance of about $200 \, \text{Å}$ between the micelles. The contribution to the self-diffusion is then $13 \times 10^{-11} \, m^2 \, s^{-1}$.

Nilsson et al. [5] found for $C_{12}E_8$ at $69.2 \, °C$ that the minimum value of the self-diffusion coefficient was $3 \times 10^{-11} \, m^2 \, s^{-1}$ at a concentration of 5–10%. The exchange contribution should be somewhat less than this value. For consistency with the pyrene migration data, we must assume that pyrene exchanges at least four times faster than the surfactant monomers on micelle collisions, an assumption which appears rather improbable.

The measured self-diffusion coefficient puts an upper limit to the exchange rate of the surfactant monomers which shows that the micelles change size slowly compared to the fluorescence decay and the migration rate of pyrene. To justify their use of the Infelta model for polydisperse samples, Zana and Weill [7] made the opposite assumption.

We have discussed, at some length, the general features of the intra- and intermicellar processes that affect the deactivation of hydrophobic probes in micelles. We will now turn to a discussion of the results obtained for the deactivation of 9-methyl-anthracene triplets by azulene or guajazulene in some different micellar systems.

Small ionic micelles

With SDS or CTA$^+$ as surfactants the fast intramicellar process was resolved only at very high micelle concentration − or when high concentrations of salt (NaCl) had been added − and low temperature, i.e. under conditions which force these micelles to grow into rods. The slow deactivation was normally a well-behaved single exponential process which, for normal migration by diffusion through the intermicellar solution should have the decay constant given by the product of k_+ and the free quencher concentration according to Eq. (4).

The concentration of free Az was calculated from the distribution constant as determined in saturation measurements, Table I. Values of k_+ from Eq.(4) are given in Fig. 1, where also k_{diff} is shown for compari-

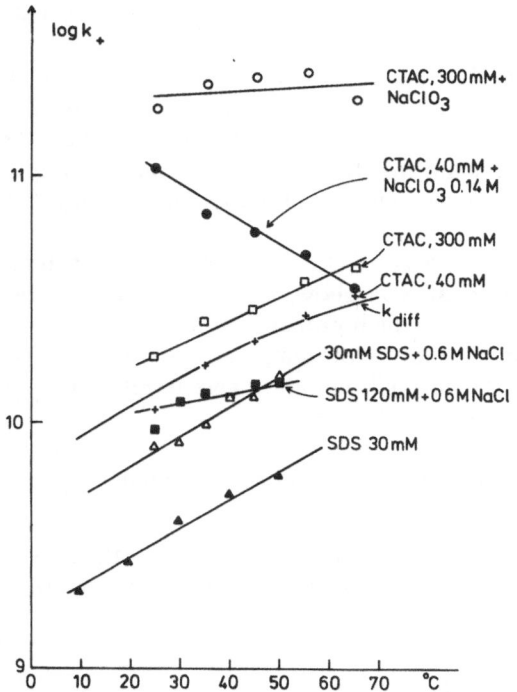

Fig. 1. The entrance rate constant k_+ ($M^{-1}s^{-1}$) for azulene into micelles in some different systems, calculated from the measured slow decay constant and the estimated concentration of free azulene, as described in the text. The composition of the solutions were as follows:
1. SDS 30 mM; Az 1.33×10^{-4} M
2. SDS 30 mM; NaCl 0.60 M; Az 1.30×10^{-4} M
3. SDS 120 mM; NaCl 0.60 M; Az 5.36×10^{-4} M
4. CTAC 40 mM; NaCl 70 mM; Az 1.42×10^{-4} M
5. CTAC 300 mM; NaCl 70 mM; 1.19×10^{-3} M
6. CTAC 40 mM, solution 4 with 140 mM NaClO$_3$ added
7. CTAC 300 mM, solution 5 with 140 mM NaClO$_3$ added. The CTAC preparation held, in reality, 85% Br$^-$ as counterion. All solutions contained about 1×10^{-4} M 9-MeA. The azulene concentrations correspond to about 0.4 Az per micelle under conditions when small micelles are present

Table 1. Distribution constant, K, of azulene between aqueous intermicellar solution and micellar pseudo-phase, as determined from saturation measurements

Surfactant	Temperature (°C)	K (M^{-1})
SDS	25	210
	50	136
SDS, 0.6 M NaCl	25	340
	50	205
CTAC, 0.07 M NaCl	25	838
	50	440
$C_{12}E_6$	25	654
	50	551

$K = C_m/[C_{aq}(C_{surf} - C_{free})]$ where index m is micellar and aq, aqueous phase, C_{surf}; total surfactant concentration in mol/l, C_{free}, concentration of free monomer.

son, calculated from Eq.(5) using parameters as for SDS in water. There is evidently large differences in k_+ between different systems. In CTA$^+$ k_+ is in general larger than in SDS, and even larger than k_{diff} in some cases. The last finding is probably an effect of that the micelles were larger than assumed in the calculation of the diffusion-controlled value, there is also some uncertainty in the estimates of the free Az concentrations. It remains to be explained, however, why there is such a large difference between SDS and CTA$^+$. Association of Az with free CTA$^+$ monomers in the intermicellar solution would promote the migration but, contrary to what was observed, it would also increase the solubility of Az in the presence of CTAC already at concentrations below the cmc.

It is notable that the temperature dependence is almost the same for k_{diff} and k_+ in most of the CTA^+ and SDS systems, suggesting that the rate limiting steps are still diffusional, but that the limit is set by the diffusion close to or across the micelle-water interface.

The k_+ values for SDS in 0.6 M NaCl at a high concentration of SDS show an exceptional temperature dependence, which probably reflects the presence of rod-like micelles at low temperatures. As discussed above, the diffusional migration gets faster when the micelles are long rods.

Cetyltrimethylammonium chloride—sodium perchlorate

After the first series of measurements, $NaClO_3$ was added to the solutions used in the study of migration

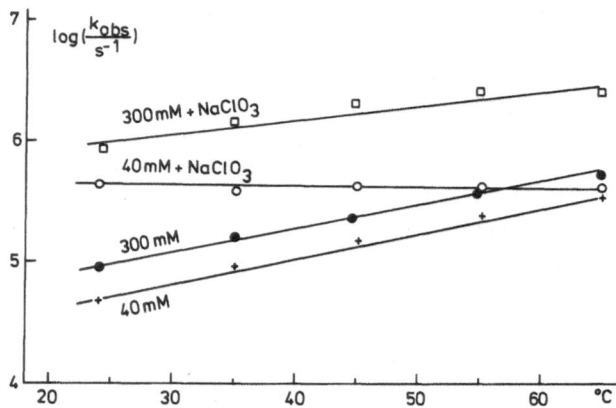

Fig. 3. Observed slow deactivation constants for the CTAC systems in Fig. 1

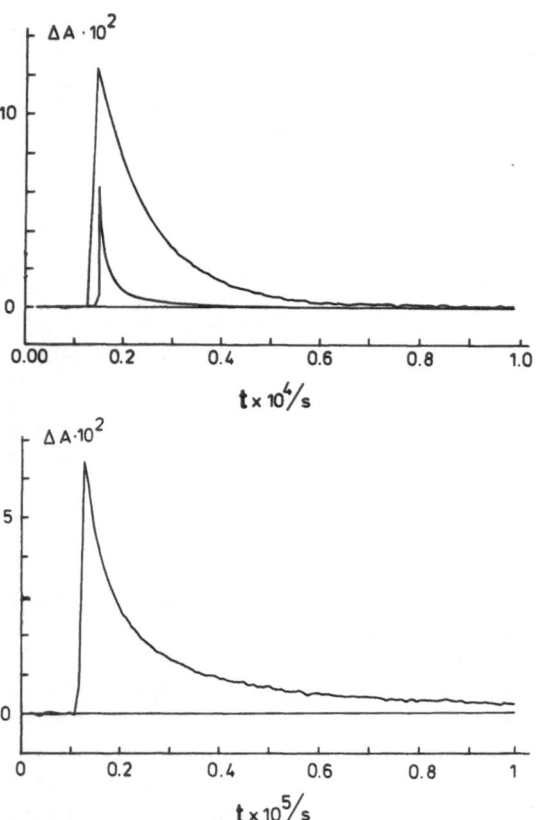

Fig. 2. Experimental decay curves which demonstrate the effect of addition of chlorate ions on the deactivation of 9-methylanthracene triplets by azulene in micellar CTAC.
A. Solutions no. 4 (the large transient) and no. 6 (small transient) from Fig. 1, at 35 °C
B. The small transient from (A) at a shorter time scale

in CTA^+. These additions produced a dramatic change in the viscosity of the solutions, indicating the formation of very extended aggregates. The change of the deactivation process was no less dramatic: as illustrated in Fig. 2, it became much faster and strongly non-exponential. The decays were analyzed in terms of two exponential processes; it should be stressed, however, that these were not well separated. We will mainly discuss the slowest decay constants, since it is in these that the effects of the migrative modes will be found. They may also be influenced by intramicellar events.

The slow decay constants are presented in Fig. 3. Values of k_+, calculated from them with Eq. (4), are presented in Fig. 1. The k_+-values are distinctively different for systems with and without chlorate, both in magnitude and in temperature dependence, showing that the final deactivation is not governed by diffusive intermicellar migration in the chlorate system.

The concentration of azulene in these measurements was chosen to give an average of 0.4 azulene molecules per CTA^+ micelle before chlorate was added. The mean occupancy number was much larger with chlorate present, perhaps by an order of magnitude. Almost no micelles would then have been without azulene, and the observed deactivation would have been entirely intramicellar. A strongly nonexponential decay extending to long times would result from the polydispersity of the solution of long rods, as observed.

A second set of experiments were performed with the concentration of azulene lower by a factor of four (Fig. 4). Before chlorate was added, the slow decay constant was decreased by the same factor, but by a factor of almost ten with chlorate present. The fast and slow decays were somewhat better separated at the

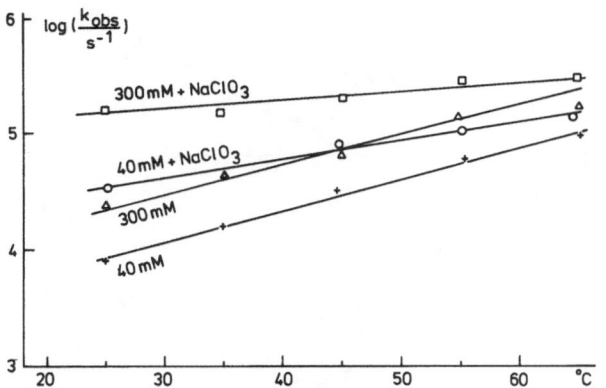

Fig. 4. Observed slow deactivation constant for CTAC systems with reduced azulene concentration. The compositions are as in Fig. 1, but with the Az concentration reduced by a factor of four

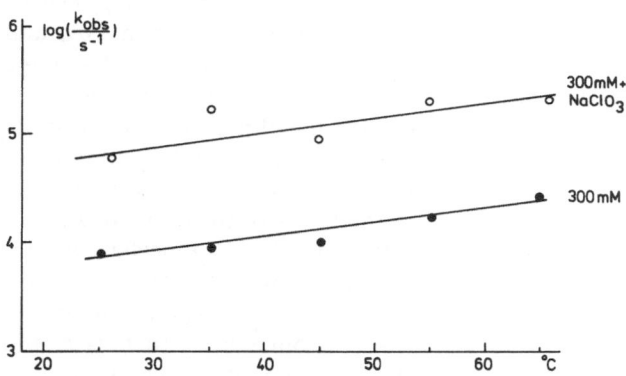

Fig. 5. Observed slow deactivation constant for CTAC systems with composition as in Fig. 1, but with guajazulene as the acceptor. The concentration of GAz is the same as that of Az in Fig. 1

low Az concentration, but the observed slow decay was probably still somewhat mixed up with the slowest part of the intramicellar process, in particular at low temperatures. At the highest temperatures, where the micelle size should be smaller, the deactivation rates approached those observed in the absence of chlorate.

From time-resolved fluorescence measurements on the excimer formation of pyrene in a system with 0.3 M CTAC, 0.07 M NaCl, 0.13 M NaClO$_3$, which is close to the composition that gave the results in Fig. 4, Malliaris et al. [6] reported an intermicellar deactivation rate constant at 25 °C of about 2×10^6 s^{-1}. The pyrene concentration was five times larger than out lowest concentration of Az. Even reduced by this factor of five the pyrene deactivation rate remains appreciably larger than that for Az. This suggests that the pyrene deactivation could be entirely intramicellar. To further substantiate this suggestion, some experiments were performed with a hydrophobic azulene derivative, guajazulene. If the slow process observed at the low quencher concentration were a normal migration process, its rate would be further reduced for a more hydrophobic compound, whereas, on the other hand, a migration in fragment carriers would remain unaffected.

The results showed that the slow deactivation was strongly suppressed, so much, in fact, that deactivation by remaining oxygene or spurious impurities became the dominant deactivation mode. Only the results from measurements at high GAz concentrations are presented in Fig. 5, therefore. A comparison with the results in Fig. 3, where Az was used at the same concentration, immediately shows that the slow deactivation was much slower with GAz, also after the

addition of chlorate, whereas the fast part of the decay was about equally fast for Az and GAz. Since the results for Az were obtained with samples containing large amounts of Br$^-$, they should not be compared in detail to those for GAz. We can compare, however, the result obtained by Malliaris et al. for pyrene, referred to above, and the slow deactivation rate with GAz. The concentration of GAz was slightly lower (by 25%) than that of pyrene; the composition of the micelle solutions were similar. The deactivation with pyrene was faster by a factor of about 20, which is clearly incompatible with a migration mechanism involving fragmentation-coagulation.

The interpretation of the results in Figs. 3 and 5 and the comparable ones for pyrene, is not clear in all respects. It seems clear that these micelle solutions contain very long rods. Results from light scattering [16], the high viscosity, and a number of methods applied to similar systems [17] all point in this direction. The estimate $N = 270$ by Malliaris et al. [6] from fluorescence measurements appears much too small. The reason is the use of the Infelta model in combination with the assumption of a fast migration. Malliaris et al. showed that a model which assumes no migration, i.e. $A_2 = k_0$ in Eq. (1), gives a bad fit with systematic deviations to the experimental data. This is to be expected for a polydisperse sample [18]. The estimated aggregation number is larger from this fit by almost a factor of two, but is probably still a rather low value of the weight averaged aggregation number [18].

With so large aggregation numbers the fraction of empty micelles, i.e. those containing only the excited probe but no deactivator, was small in the solutions.

Only a very small fraction of the decay may then be ascribed to a pure intermicellar deactivation. GAz and pyrene are both very hydrophobic probes. The large difference observed between the two are probably just caused by the limited lifetime of pyrene which prevents the detection of a slow decay with small amplitude. In the deactivation of 9-MeA by GAz the slow decay had an amplitude of less than 10% of the total. We cannot decide whether it was entirely intramicellar or had migrative contributions. The much faster deactivation with Az, compared to GAz, suggests migration through an aqueous environment in that case: the migration of the more hydrophilic azulene molecules from one long rod to its neighbor is so fast that it competes effectively with diffusion along the rods, in those sections where the excited probe was created far from any deactivator in the same micelle. The cross over from one rod to another may become particularly effective if the rods approach each other at some "entanglement points" [19].

Nonionic surfactants

For $C_{12}E_6$ fluorescence quenching data on the aggregation numbers are available for temperatures up to the cloudpoint [7, 20]. The values given by Zana and Weill [7] were based on the Infelta equation and the assumption of rapid migration of pyrene, which, as discussed above, gives too low aggregation numbers. It was clearly shown that a growth occurs, also for $C_{12}E_8$, starting at temperatures 25 °C below the cloud point [7]. We have studied these systems extensively; here a straightforward comparison will be made of the results for two solutions of $C_{12}E_6$ and $C_{12}E_8$, respectively, with the same molar concentrations, and the same concentration of azulene.

In both cases (Fig. 6) only the slow migration mode was resolved at the lowest temperatures. At about 25 °C below the cloudpoints, the micelles started to grow so that the intramicellar mode had sufficient amplitude and was sufficiently slow to be resolved. For $C_{12}E_6$ this occurred just above 20 °C; simultaneously the rate of the migration process starts to rise above the still normal values for $C_{12}E_8$ (which closely show the same temperature dependence as observed for SDS). The separation between the fast and slow process is quite good — about an order of magnitude — but, perhaps, at the highest temperatures.

Since aggregation numbers are available for $C_{12}E_6$ it is possible to make a closer consistency test of the results (Table 2). The average number of azulene molecules per micelle is calculated (disregarding the

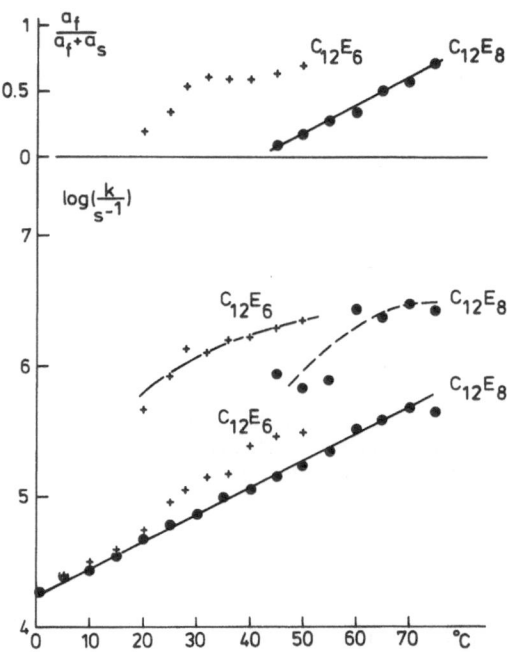

Fig. 6. Fast and slow decay constants, and the relative amplitude of the fast process for $C_{12}E_6$ and $C_{12}E_8$. Compositions: $C_{12}E_6$ 2% w/w; Az 1.33×10^{-4} M $C_{12}E_8$ 2.35% w/w; Az 1.30×10^{-4} M

Table 2. $C_{12}E_6$ micelles at various temperatures. Weight average aggregation numbers and width of the micelle size distribution. The mean number of azulene molecules per micelle, the relative amplitude of the slow process, decay constant of the fast process, and deduced intramicellar rate constant for triplet energy transfer to azulene.

Temp. °C	$\langle N_s \rangle_w$ [a]	σ [a]	n	$P(0)$ [b]	a_s [c]	$k_f \times 10^{-6}$ (s^{-1}) [d]	$k_q \times 10^{-6}$ (s^{-1}) [e]
17	150	50	0.47	0.63	0.9	–	–
25	250	80	0.74	0.48	0.6	1.0	1.3
35	370	100	1.12	0.33	0.35	1.2	1.1
45	1200	600	3.6	0.03	0.3	1.5	0.4

a: From Ref. [20]
b: Fraction micelles without quenchers, from $P(0) = \exp(-n)$
c: Relative amplitude of slow process (Fig. 6)
d: Observed decay constant for the fast process
e: Intramicellar quenching constant from $k_q = k_f/n$

polydispersity) and used to estimate the fraction of "empty" micelles. This fraction changes much the same as the observed amplitude of the slow process but at the highest temperature, indicating that the slow process is intermicellar at low and intermediate temperatures, but seemingly entirely intramicellar at the highest temperature. If we further assume that the observed decay rate of the fast process is $k_q n$, the initial rate according to Eq. (3), an estimated of k_q is obtained (probably an underestimate since the entire initial portion is not resolved). This estimate yields a value of 4×10^5 s^{-1} at the highest temperature, which is just sufficient to ascribe the rate of the slow process in this case to deactivation in micelles with one quencher; the amplitude, however, is too large by a factor of about three. Thus, although not entirely consistent at the highest temperatures, the interpretation is sensible, and we proceed to discuss the increased migration rate at intermediate temperatures, where the growth in size is moderate. We assume that the observed decay constant is the sum of the normal migration contribution, $k_+ [Az]_{aq}$ and an exchange on micelle collisions, $k_{ex} n$. The values observed for $C_{12}E_8$ are taken to represent the normal contribution, giving $k_{ex} n = 3 \times 10^4$ s^{-1} at 25°C and 8×10^4 s^{-1} at 35°C for $C_{12}E_6$. Comparing this with the expected frequency of micelle collisions, the exchange probability per collision is obtained as 0.035 at 25°C and 0.13 at 35°C.

With an average distance between the micelles of about 200 Å an exchange frequency of 5×10^4 s^{-1} would give a contribution to the self-diffusion, Eq. (6), of about 3×10^{-12} m^2 s^{-1}, which is not too large when compared to the minimum values of 2×10^{-12} m^2 s^{-1} and 10×10^{-12} m^2 s^{-1}, respectively, obtained by Nilsson et al. [5] for the total self-diffusion coefficients of $C_{12}E_5$ and $C_{12}E_8$ at a temperature of 25°C below the respective cloud point.

These results are thus compatible with an exchange of hydrophobic solutes on micelle collisions as suggested by Zana and Weill, although the process is much slower than they proposed.

Conclusions

In all the systems carefully studied residence times in the micelles exceeding a microsecond were observed for hydrophobic solutes. The very fast migration claimed to occur in the same micelle systems under similar conditions, is most probably a misinterpretation of intramicellar events. For small ionic micelles the migration of azulene is substantially slower than expected for a diffusion-controlled process but have the same temperature dependence. At high counterion concentration, and in particular when the counterion is strongly bound as the chlorate ion with CTA$^+$, the migration rate increases, which is also the case for nonionic micelles under conditions where these are large and interact strongly. The mechanisms responsible for the increase are not completely clear: exchange on micelle collisions seems important in some cases, in others there is a dependence on the hydrophobicity of the migrating species that is not fully understood.

Acknowledgement

We are indebted to Docent Å. Ohlin for Br$^-$ analysis of the CTAC-CTAB samples. The work was supported by the Swedish Natural Science Research Council.

References

1. Almgren M, Grieser F, Thomas K (1979) J Am Chem Soc 101:279
2. Lessner E, Teubner M, Kahlweit M (1981) J Phys Chem 85:3167
3. Kahlweit M (1981) Pure Appl Chem 53:2069
4. Kahlweit M (1982) J Colloid Interface Sci 90:92
5. Nilsson P-G, Wennerström H, Lindman B (1983) J Phys Chem 87:1377
6. Malliaris A, Lang J, Zana R (1986) J Phys Chem 90:655
7. Zana R, Weill C (1985) J Phys Lett 46:L-953
8. Infelta PP, Grätzel M, Thomas JK (1974) J Phys Chem 78:190
9. Almgren M, Löfroth J-E, van Stam J (1986) J Phys Chem 90:4431
10. Almgren M, Löfroth J-E (1981) J Colloid Interface Sci 81:486
11. Malliaris A, Le Moigne J, Sturm J, Zana R (1985) J Phys Chem 89:2709
12. Malliaris A, Lang J, Zana R (1986) J Chem Soc, Faraday Trans I 82:109
13. Van der Auweraer M, Dederen J, Geladé E, De Schryver F (1980) J Phys Chem 74:1110
14. Sano H, Tachiya M (1981) J Phys Chem 75:2870
15. Herrmann C-U, Kahlweit M (1980) J Phys Chem 84:1536
16. Porte G, Appell J (1984) In: Mittal KL, Lindman B, eds. Surfactants in Solution. Plenum Press, New York, Vol 2, p 805
17. Bayer O, Hoffmann H, Ulbricht W, Thurn H (1986) Adv Colloid Interface Sci 26:177
18. Almgren M, Löfroth JE (1982) J Chem Phys 76:2734
19. Candau SJ, Hirsch E, Zana R (1985) J Colloidal Interface Sci 105:521
20. Löfroth J-E, Almgren M (1984) In: Mittal KL, Lindman B, eds. Surfactants in Solution. Plenum Press, New York, Vol I, p 627

Received September 30, 1986;
accepted January 20, 1987

Authors' address:

Mats Almgren
The Institute of Physical Chemistry
University of Uppsala
P.O. Box 532
S-75121 Uppsala, Sweden

Progress in Colloid & Polymer Science

Progr Colloid & Polymer Sci 74:64–68 (1987)

A moving contact line: Further studies of "Haines' jumps"

M. A. Cohen Stuart and A. M. Cazabat

Collège de France, Physique de la matière condensée, Paris, France

Abstract: When a liquid containing surface active molecules advances on a hydrophilic solid surface, specific instabilities are observed. They are due to the diffusion of the surface active molecules which adsorb on the surface, increasing the contact angle. These instabilities appear as waves travelling along the contact line and are visible only in a limited range of liquid edge velocities.

Key words: Wetting, adsorption, contact angle, surfactants, monolayers

Introduction

Slowly advancing contact lines often show a special type of motion, where the line jumps between stable positions. Such effects were first reported by Haines [1] and seem to occur quite frequently (though mostly involuntarily) during formation of Langmuir-Blodgett layers [2]. In some texts, the same phenomenon is denoted as "stick-slip" behaviour [3] probably in analogy with a phenomenon well known for solid/solid contacts. Mason and coworkers, notably Bayramli et al. [4], observed a saw-tooth-like periodicity in the capillary force of water wetting a high energy surface. They concluded that the phenomenon (referred to as "Haines' jumps") was due to traces of surface active impurities present in the liquid phase*), which are deposited just in front of the line.

Indeed, this explanation is consistent with all experimental data on Haines' jumps. It is well known that high energy surfaces such as oxides and metals are well wetted by a large variety of liquids**). However, a few amphiphilic molecules may effectively reduce the wettability as, for example, seen by contact angles [6]. Since a complete monolayer usually contains no more than 10^{-10} mol cm^{-2}, the quantities of surface active material involved are very small. Hence, extreme care is needed in studies of wetting of high energy surfaces [7, 8], especially in dealing with water.

We have studied Haines' jumps for organic liquids (hexadecane, trioctylamine) on glass, taking a plane geometry rather than a cylindrical one, as in Ref. [4]. This allowed us to easily make optical observations by means of a projection arrangement. Also, complications such as the coupling of a (closed) contact line with itself are avoided.

Experimental

Apparatus

Microscope glass slides could be immersed slowly under a variable angle a with the liquid surface plane. The velocity of immersion could be varied continuously over four orders of magnitude between $\sim 10^{-2}$ and $\sim 10^2$ µm s^{-1}. The contact line was illuminated with white light, and video recordings or photographs could be taken. The equipment was mounted in such a way as to minimize vibrations.

Materials

Clean glass surfaces were prepared by washing thoroughly with a detergent, soaking for about 1 min in hot sulfochromic acid and rinsing with three-times distilled water and methanol. The final cleaning consisted of UV illumination in a stainless steel cell, filled with oxygen for at least 30 min [9].

Liquids were hexadecane (SDS) and trioctylamine (Merck), as provided by the manufacturer.

---

*) We thank Dr. T. G. M. van de Ven for bringing reference [4] to our attention. In fact, we had drawn the same conclusion without being aware of the work described in ref. [4].

**) We note however that Haines' jumps are perhaps not limited to high energy surfaces. Blake [5] gives an example for a liquid/liquid/solid three phase line on hydrophobic glass.

Results and discussion

First of all, we noted that the contact line never jumps as a whole; rather a local instability causes the line to protrude into the zone of non-wetted surface ('nucleation') whereafter the deformation propagates along the contact line as a sort of wave, without changing shape. It may be tempting to understand the properties of that wave in terms of the well known solitary wave (soliton), but we should emphasize here that such an analogy is poorly defined, because the equations of motion for the wetting wave have not yet been studied. The velocity of propagation was of the order of $1 \, \text{cm s}^{-1}$, but was rather variable, probably due to subtle variations in surface properties of the solid along the contact line. Since we did not observe systematic effects as a function of any experimentally controlled variable, we did not pursue this point. An example of a travelling wave is shown in Fig. 1, where we also indicate the size of the jump A.

A systematic variation was observed for A as a function of the angle of immersion α, as can be seen in Fig. 2. A varies as $1/\sin \alpha$. (Actually, a mean jump size was measured by counting the number of jumps for a given length of path travelled by the slide.) This suggests a well-defined condition on the angles θ_1 and θ_2 between which the line jumps.

For two line positions fixed by contact angles θ_1 and θ_2 on a surface tilted by α degrees with respect to the horizontal, one calculates the distance between them (see Fig. 2, inset) as

$$A = 4(\kappa \sin \alpha)^{-1} \sin \left(\frac{\theta_1 - \theta_2}{4} \right) \cos \left(\frac{2\alpha - \theta_1 - \theta_2}{4} \right) \tag{1}$$

where $\kappa^{-1} = (\gamma/\rho g)^{1/2}$ is the capillary length fixed by the surface tension γ, the density ρ and the gravitational acceleration g. For angles α up to $35°$ and θ_1, θ_2 not very large the last factor in Eq. (1) is about unity, so that a linearity in $(\sin \alpha)^{-1}$ is consistent with a constant difference between θ_1 and θ_2. Apparently, the line jumps between two stable positions, each one corresponding to a well-defined angle. Taking κ^{-1} to be equal to 1.7 mm and 1.9 mm for hexadecane (HD) and trioctylamine (TOA), respectively, we find jumps

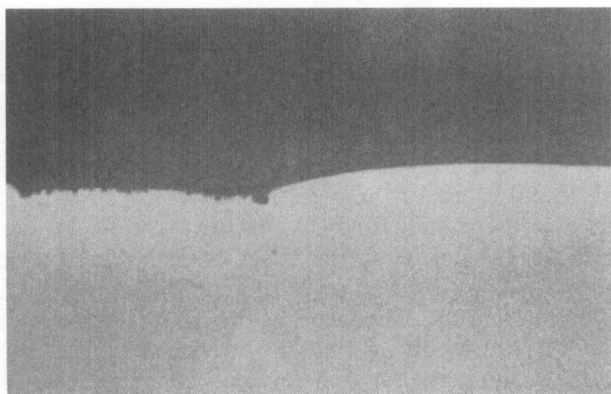

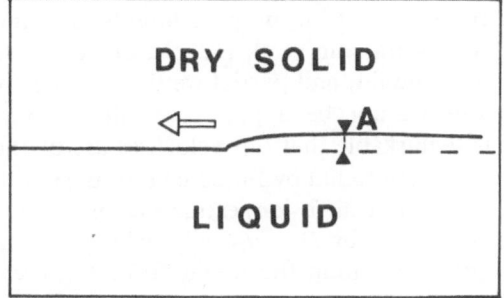

Fig. 1. Photograph of a Haines' jump, and explanatory diagram

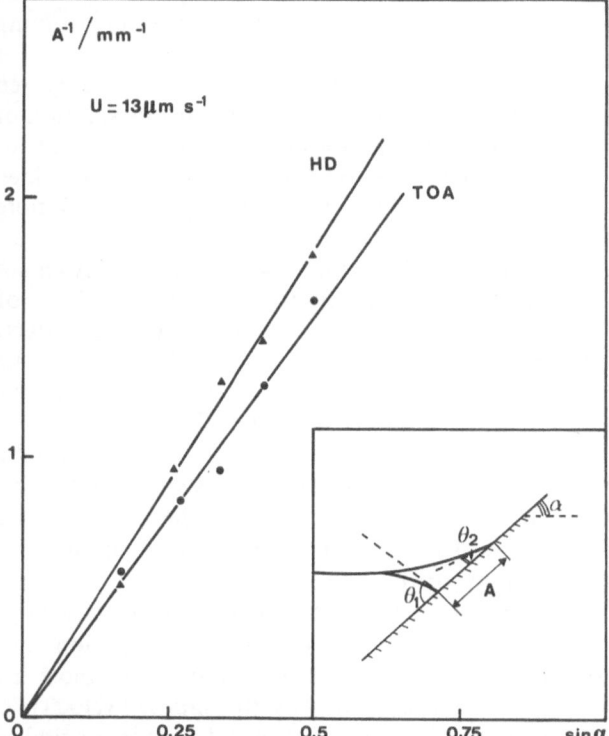

Fig. 2. Reciprocal size of jumps, A^{-1}, as a function $\sin \alpha$, where α is the dipping angle. (▲) hexadecane; (●) trioctylamine

(i.e. $\theta_1 - \theta_2$) of the order of 10°, which would agree with jumps between 20° and 10° for hexadecane and between 30° and 20° for TOA. This agrees with measurements of advancing and receding contact angles on the same systems.

These observations seem compatible with the following mechanism. On first contact, the contact angle tends to assume the value corresponding to the solid/pure liquid system. However, our liquids are impure and contain an (unknown) concentration of molecules which can adsorb on the solid. (For the moment we will not speculate what kind of molecules.) Hence, after a lapse of time, the surface will be covered with a layer of impurity molecules and the corresponding contact angle will be increased.

If the solid surface did not move with respect to the liquid reservoir, one would expect the line to recede spontaneously after a while. However, this recession is compensated largely by the motion of the slide being immersed, so that the line remains stationary with respect to the solid surface. As soon as the position of the line corresponds to the upper critical angle θ_1 (for the surface with adsorbed molecules) any further displacement must result in forward motion with respect to the solid surface. Since this brings the liquid into contact with the clean surface, a smaller contact angle, θ_2, is imposed, causing the contact line to jump to the corresponding position.

In order to test this mechanism, we also compared pure triply-distilled water with an aqueous solution of CTAB. The water did not jump, but the solution produced very clear jumps. This corroborates the idea that the effect is due to an adsorbing solute. We note in passing that because, in these experiments, the surface area/solution volume ratio is extremely small ($\sim 10 \text{ m}^{-1}$) concentrations of the order of $10^{-7} - 10^{-8}$ mol/l are sufficient to produce the effect. For the moment we cannot definitely rule out the possibility of impurity transport through the vapour phase, but we tend to think that such transport is very sensitive to uncontrolled convective processes in the vapour phase, and therefore inconsistent with the nice reproducible result we had. (Some other results, to be discussed below, also do not fit in with the idea of vapour transport.)

Since the above interpretation clearly involves a dynamic situation (adsorption kinetics, diffusion and convection of adsorbate, liquid motion) the most interesting variable is, obviously, the imposed velocity u. We noted, as did Bayramli et al. [4], that in a suitable velocity range the line advances only by jumps, so that after n jumps of size A the line has travelled a distance $n.A$. However, if the velocity is reduced, the number of

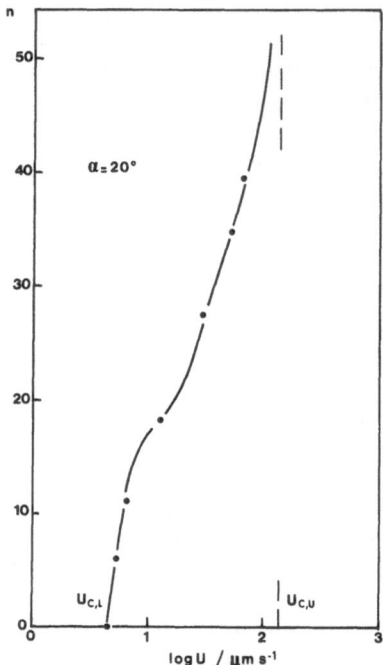

Fig. 3. Number of jump nucleations as a function of $\log u$ (u is the dipping velocity) for trioctylamine

jumps decreases rather abruptly (their size remaining the same). We observed this for TOA (Fig. 3). Between jumps, the contact line moves smoothly for periods which increase with decreasing velocity, to disappear altogether at a lower critical velocity $u_{C,L}$. Bayramli et al. found values of the order of 10^{-2} μm s^{-1} for $u_{C,L}$, whereas we found, for TOA, 4.4 μm s^{-1} (Fig. 3). For HD we did not reach the lower critical velocity; we suppose it is at least an order of magnitude lower than the one for TOA. The fact that the contact line moves smoothly must mean that the surface ahead of the contact line is already well covered with surface active molecules. Hence these molecules spread on the dry surface fast enough to meet this condition. Thin films ahead of the contact line ("precursor films") are expected in complete wetting by pure liquids [11] and were detected by Bascom et al. [12]. In our case, we have an impure liquid, and partial wetting so, strictly speaking, this cannot be a precursor film. Nevertheless, it is remarkable that our velocities are of the same order as those found by Bascom et al. Bayramli, reasoning in terms of diffusion, extracts a surface diffusion coefficient D_s by $D_s = u_{C,L} \cdot Z_C$, where Z_C is a critical length, separating the liquid front from the clean surface.

The idea is that a sufficiently wide zone covered by surfactant just ahead of the contact line will efficiently block it and reduce the 'nucleation' probability to zero. It is not clear how wide such a zone has to be; in Ref. [4] a width of 1 nm is proposed, which is probably reasonable in view of thermal fluctuations of the line. This would lead to $D_s = 4.4 \times 10^{-15}\,\mathrm{m^2\,s^{-1}}$ for TOA. At this rate, a film several microns wide could develop within a few hours. The fact that molecules can easily spread over a surface, without passing through a vapour phase, has important consequences for the phenomenon of complete wetting. In a recent study on complete wetting [13] we came to the conclusion that, on theoretical grounds, one expects a precursor film to be usually molecularly thin. This casts some doubts on the hydrodynamic calculations pertaining to precursor films [11] and we might suggest here that a diffusion mechanism seems more appropriate.

An entirely new observation is that there is also an upper critical velocity $u_{c,v}$. This is immediately clear from Figs. 3 and 4. As the dipping speed approaches $u_{c,v}$ the number of spontaneous nucleations increases rapidly, so that a certain length of contact line carries many waves at the same time. Where two waves travelling towards each other meet, they annihilate to the ex-

tent that their sizes are equal. Hence, since there are now many waves, the average lifetime of a wave becomes shorter. At $u_{c,v}$ this lifetime vanishes altogether and separate waves are no longer detectable. Instead, the contact line shows noise-like perturbations.

The increase in imposed line velocity has two consequences. Firstly, the viscous force on the contact line increases: a purely hydrodynamic effect. Secondly, the rate of deposition (adsorption) of the surfactant molecules (which in our case of an apolar solvent, are present in the solution rather than at the liquid/gas interface) may be too slow to keep up with the line velocity, so that no barrier builds up. However, the movement of the liquid over the solid surface gives rise to an important convective transport, so that the latter effect seems rather unlikely. Moreover, our experimental data show that the ratio between the upper critical velocities for our two liquids ($u_{c,v}$(HD)/$u_{c,v}$ (TOA) = 3) exactly equals the inverse ratio of their viscosities (η (TOA)/η (HD) = 9.8/3.3 = 3). Hence, we tend to interpret the upper critical velocity as purely hydrodynamically determined.

Conclusions

We have independently confirmed the nature of so called "Haines' jumps" often observed on liquid fronts advancing over high energy surfaces. In addition, we have studied the effect of tilting the solid surface with respect to the horizontal; a mechanism whereby the line jumps between two fixed contact angles explains the results very well. Finally, we have identified not only a lower critical velocity, wher surface diffusion becomes important, but also an upper critical velocity due to hydrodynamic effects. Experiments in which we vary the nature and concentration of the surfactant are now in progress.

Acknowledgements

We acknowledge the cooperation of A. Bouillault, who assisted with most of the experiments. We also benefitted greatly by discussions with P. Fromherz and K. Mysels.

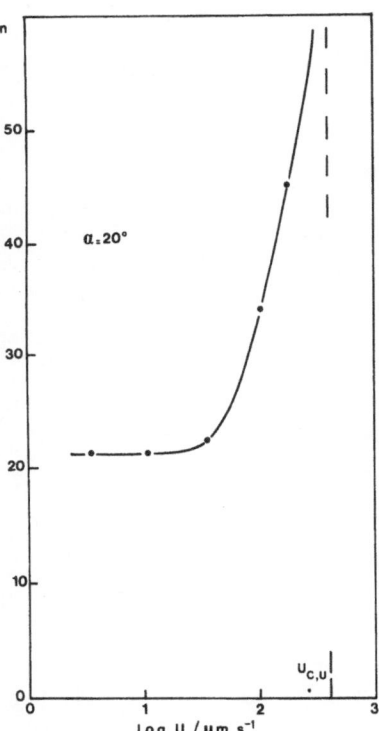

Fig. 4. Number of jump nucleations as a function of log u (u is the dipping velocity) for hexadecane

References

1. Haines WG (1930) J Agric Sci 20:97
2. Formherz P, private communication; Richardson RM, Buhaenk M, private communication
3. Dussan EB (1979) Ann Rev Fluid Mech 11:371
4. Bayramli E, van de Ven TGM, Mason SG (1981) Colloids and Surfaces 3:131

5. Blake TD, PhD Thesis (1968) Univ Bristol, England
6. Hi W, Gu T (1985) Colloid Polym Sci 263:1041
7. Yekta-Fard M, Ponter AB (1984) Phys Chem Liq 89
8. Bewig KW, Zisman WA (1961) Adv Chem Ser 33:100
9. Vig JR (1985) J Vac Sci Technol A3(3):1027
10. Jansons KM (1985) J Fluid Mech 154:1
11. De Gennes PG (1985) Rev Mod Phys 57:827
12. Bascom WD, Cottington RL, Singleterry CR (1964) Adv Chem, Contact Angle, Wettability and Adhesion 43:355
13. Cazabat AM, Cohen Stuart MA, Colloid Polym Sci (this issue, in press)

Received September 29, 1986;
accepted December 4, 1986

Authors' address:

A. M. Cazabat
Collège de France
11 Place Marcelin Berthelot
F-75231 Paris Cedex 05, France

Progress in Colloid & Polymer Science   Progr Colloid & Polymer Sci 74:69–75 (1987)

Dynamics of wetting on smooth and rough surfaces

A. M. Cazabat and M. A. Cohen Stuart

Collège de France Physique de la matière condensée, Paris, France

Abstract: The rate of spreading of non-volatile liquids on smooth and on rough surfaces was investigated. The radius of the wetted spot was found to agree with recently proposed scaling laws ($t^{1/10}$ for capillarity driven and $t^{1/8}$ for gravity driven spreading) when the surface was smooth. However, the crossover between these regimes was not observed at a constant value of the radius. Rough surfaces exhibited at least four spreading regimes which could be rationalized in terms of a macroscopic contact angle and macroscopic deviations thereof. An interesting feature is the appearance of a wetted rim around the central drop. This rim follows a diffuce spreading law ($t^{1/2}$). Partially wetting liquids followed the same dynamics provided the spreading conditions was fulfilled.

Key words: Wetting dynamics, smooth and rough surface, silicone oils

Introduction

Because of their practical importance, wetting phenomena have raised many questions for a long time. Moreover, the subject has been more recently renewed, both from the theoretical [1–6] and the experimental [3, 7–8] point of view.

A very simple analysis of the dynamics of complete wetting and the role of the precursor film has been drawn up by De Gennes [6].

The aim of the present work is to check some of these predictions [6] for complete wetting on smooth surfaces. We have also used rough surfaces with controlled roughness. Similar systems were previously studied by Mason and coworkers in non-wetting situations [8–11].

After a short summary of De Gennes analysis (I) we shall present and discuss our experimental results for drops spreading on smooth surfaces (II). Then, the behavior of a drop spreading on a rough surface is briefly described. A simple model is proposed and compared to experiment (III). A comparison with more elaborated models follows (IV). Finally, some experiments with non-wetting liquids are presented (V).

I. Theoretical predictions for spreading on smooth surfaces

Let us consider a drop of a non-volatile liquid ("dry" spreading) which spreads on a smooth horizontal surface, completely wetted by the liquid.

During the spreading, the volume Ω of the drop is conserved. The rate of spreading is characterized by the time dependence $R(t)$ of the wetted spot, or by the dynamic contact angle $\theta(t)$.

In the case of complete wetting, the spreading parameter S

$$S = \gamma_{SG} - \gamma - \gamma_{LS} \tag{1}$$

is positive. Here γ_{SG} is the interfacial tension between the solid and the gas (air), γ the surface tension of the liquid, γ_{LS} the interfacial tension between solid and liquid. A very thin precursor film always precedes the visible droplet [6] and plays a major role in the balance of the forces which control the spreading process.

The driving forces for spreading are capillary and gravity forces. They are balanced by viscous friction forces. Per unit length of contact line, the capillary forces can be written as [6]

$$S + \gamma(1 - \cos\theta) \quad \text{or, for small } \theta$$

$$S + \tfrac{1}{2}\gamma\theta^2. \tag{2}$$

The gravity forces are approximately

$$\sim \rho g R^2 \theta^2 \tag{3}$$

where ρ is the liquid density and g the gravitational acceleration.

The friction forces are calculated using a simple model of a wedge of liquid moving with constant velocity dR/dt. When the spreading parameter is large, the precursor film is well developed and the friction forces in the cap and in the film can be calculated separately. Within these approximations*), the friction in the cap is

$$\sim \eta\,\theta^{-1}\frac{dR}{dt} \tag{4}$$

where η is the liquid viscosity. The friction on the film is just S [6].

As a consequence, the S term in the capillary force is just balanced by friction in the precursor film. So the dynamics of the cap (which is the visible part of the droplet) is obtained by balancing the remaining driving forces

$$\sim \gamma\theta^2 \qquad \text{for capillarity}$$

$$\sim \rho g R^2 \theta^2 \qquad \text{for gravity}$$

by the viscous force

$$\sim \eta\,\theta^{-1}\frac{dR}{dt}$$

with the condition of constant volume Ω, say

$$R^3\theta = \text{cst}. \tag{5}$$

Two limiting situations may occur:

$$\text{if} \quad \gamma\theta^2 \gg \rho g R^2 \theta^2 \quad \text{or} \quad R \ll \left(\frac{\gamma}{\rho g}\right)^{1/2}$$

the capillarity is the dominant term. Balancing it by

---

*) A rigorous deduction of (3) and (4) is given in Ref. [6].

friction one gets the "capillarity regime" extensively studied by Tanner [7]

$$R(t) \sim \Omega^{3/10}\left(\frac{\gamma t}{\eta}\right)^{1/10} \tag{6}$$

if R is much larger than the capillary length $(\gamma/\rho g)^{1/2}$ the gravity is dominant [4]

$$R(t) \sim \Omega^{3/8}\left(\frac{\rho g t}{\eta}\right)^{1/8}. \tag{7}$$

The transition between these limits is expected to occur for a given value R^* of the order of the capillary length

$$R^* \sim \left(\frac{\gamma}{\rho g}\right)^{1/2}. \tag{8}$$

In view of the preceding predictions, we decided to check the expected power laws, and to look for the crossover between them.

II. Experimental results from smooth surfaces: Discussion

1. Materials and methods

We have studied the spreading of silicone oil drops (PDMS: polydimethylsiloxane) on smooth hydrophilic or hydrophobic glass surfaces. Silicone was chosen because it wets glass well and can be obtained within a large range of viscosities, with constant surface tension γ.

We used oils with viscosities of between 0.05 and 1 Pa·s, to avoid both evaporation and specific polymer [6] behavior. The drop volume varied between 0.35 and 40 μl, the thickness at the center of the drop was always larger than 40 μm.

With non-polymeric liquids, the edge of the drop can be easily observed by simple optical imaging. The radius $R(t)$ is then measured versus time.

2. Experimental results

After a short transient situation (<30 s) with more complicated inertial effects [12], the experimental results become strictly reproducible.

An example is given in Fig. 1. Straight lines with expected slopes 1/10 and 1/8 have been drawn through the experimental points. Both capillarity and gravity regimes are successively observed [13].

Plots of log R versus log Ω at constant time inside the gravity or the capillarity regime also give a good agreement with the expected exponents 3/8 and 3/10, respectively. Finally, the expected exponents of the viscosity were also found experimentally [13].

Therefore, we conclude that all the predicted exponents for the variation of R with time t, drop volume Ω and viscosity η have been observed, both in the capillarity and in the gravity limits.

However, the crossover between these limits is not observed at constant R. It is located around a line which roughly corresponds to [13]

$$R\left(\frac{dR}{dt}\right)^{-2/3} = \text{cst}. \tag{9}$$

This constant value does not depend on η, but changes with the value of the spreading parameter. Actually, we repeated the spreading experiments on hydrophobic glass surfaces, obtained by silane grafting. The results are reported in Fig. 2, which shows a shift of the crossover towards longer times in the case of hydrophobic surfaces.

3. Discussion

Obviously, the dynamics of spreading in limit situations $R \gg R^*$ or $R \ll R^*$ are well described theoretically.

It is very strong confirmation of the fact that the cap behavior does not depend on the spreading parameter S, which has not always been recognized in the past [12]. This confirmation was needed, because some hypotheses in the theoretical analyses are rather questionable.

Firstly, the precursor film is treated hydrodynamically. It is one of the results of De Gennes' approach [6] to give an expression of the thickness and the length of this precursor. It appears that usually (and it is strictly true for dominating Van der Waals forces) the film thickness is in the molecular range. Hence, in the dynamics of the film, other phenomena, such as surface diffusion, must play a role. From this point of view, complete wetting may have much in common with surface diffusion of impurities from an impure liquid in contact with a solid [14].

Second, the dissipation in the film has been calculated using a stationary approximation $[(dR/dt) = \text{cst}]$ and in the large S limit. It was not obvious that the result would be correct up to terms of the order of $\gamma\theta^2$, i.e. much smaller than S. This point is probably related to the observed anomalous location of the crossover between the limit situations Eq. (9) and has to be clarified.

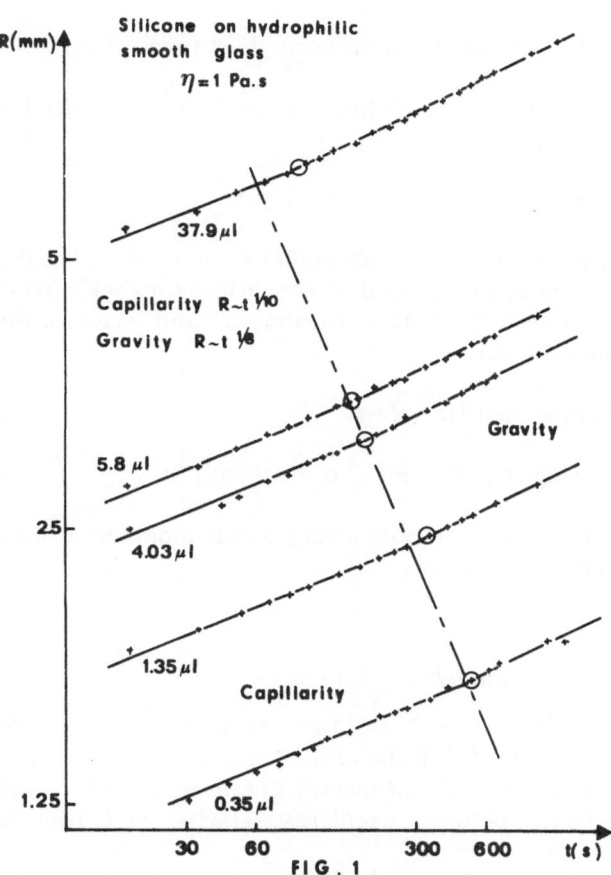

Fig. 1. Logarithmic plot of wetted spot radius versus time. Circles and dashed line: tentative location of the transition between capillarity and gravity ranges. This lines *does not* depend on the viscosity

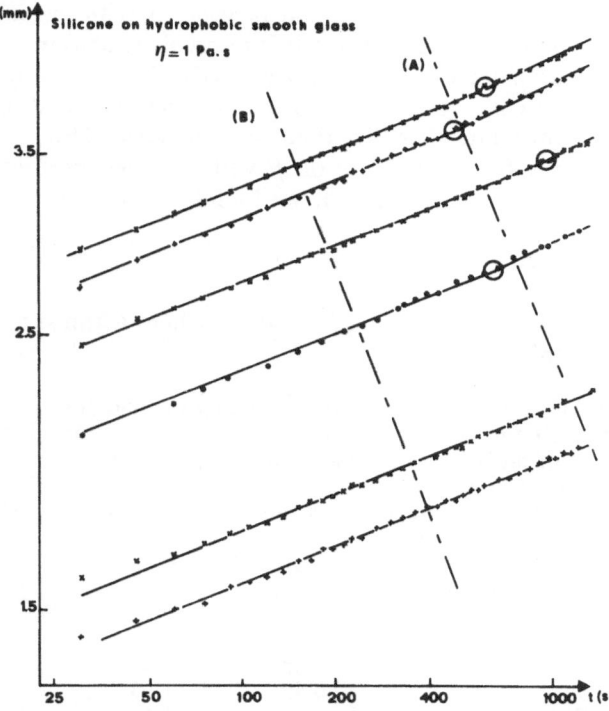

Fig. 2. Same as Fig. 1, but for hydrophobic glass. The transition of Fig. 1 is recalled as (B)

As we found these results rather encouraging, we performed the same experiments on rough surfaces.

III. Spreading on a rough surface

The contact angle hysteresis on rough surfaces has been extensively studied by Mason and coworkers [8–11]. The roughness was due to parallel, radial or spiral grooves or produced by bead blasting.

Our work concerns the dynamics of complete wetting of rough surfaces. The roughness was obtained by sand blasting (for large roughnesses) or by using different grades of abrasive powder for finer roughness. The samples were prepared and characterized at the workshop of the Ecole Supérieure d'Optique (Orsay-France). Both the amplitude h and mean quadratic variation Δh are given in Table 1.

1. Experimental observations [13]

The successive stages of the spreading of a drop on a rough surface are summarized in Fig. 3. In practice, not all are generally observed. The first ones (small times) are lacking on very rough surfaces and the last ones on almost smooth surfaces, with the time of experiment (between 15 s and 2 weeks).

First, the drop spreads as a whole (I). After some time a "foot" of liquid finds "navigable channels" [11] into the roughness and precedes a "cap" which feeds the foot (II). When this reservoir is consumed, the drop again spreads as a whole (III). Finally a slowing down is observed when the thickness of the drop becomes much smaller than the roughness (IV).

Only the dependence on R with time has been investigated. During step I, we found

$$R \sim t^{1/8}$$

excepted fot the smoothest glass, where a transition towards

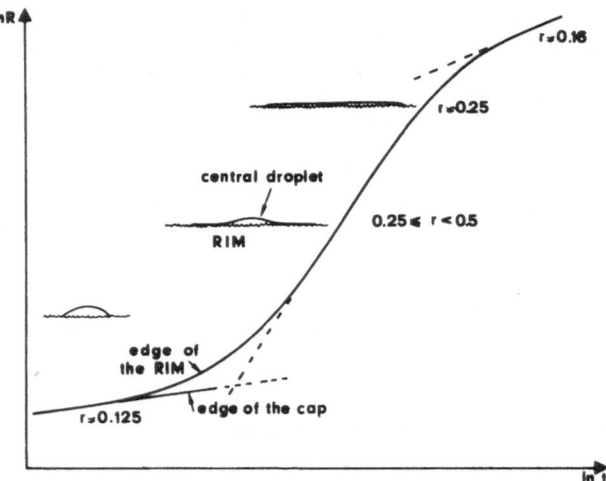

Fig. 3. Schematic logarithmic plot of wetted spot radius versus time for rough surfaces. The successive apparent power laws $R \sim t^r$ are indicated, together with the drop behavior

$R \sim t^{1/10}$ seems to occur at short times.

During step II we found an unchanged velocity for the cap

$$R_{\text{cap}} \sim t^{1/18} \, .$$

The edge of the rim also follows a power law, but the exponent does depend on the surface roughness. (It is close to 0.25 for large roughnesses and increases for small ones.)

During step III $R \sim t^{0.25}$

During step IV $R \sim t^x$ $0.16 < x < 0.17$.

We will now propose a very simple model to explain these observations.

2. Naive model

As for smooth surfaces, we assume a complete balancing of S in the microscopic precursor film [6]. The dynamics of the macroscopic drop depends on the balance between capillarity, gravity, and friction forces.

Now, the dynamic contact angle is a local variable. Formally we can write:

$$\theta^2(t) = \theta_s^2(t) + \delta^2(t) \, .$$

Table 1. Roughness amplitude h and mean quadratic variation Δh for various rough glass samples. The characteristic rate of diffusive spreading (regime II), A, is also given

Sample	h (μm)	Δh (μm)	$A = \eta D$ ($\times 10^{11}$ N)
I	50	12.0	–
III	29	6.5	89
V	18	3.6	40
VI	4.7	0.71	8
XIV	3.4	0.48	4.2
XVI	2.8	0.49	1.65
XII	2:4	0.37	1.1
IX	2	0.33	0.5

θ_s is the contact angle on the smooth surface, δ^2 a positive mean contribution of the roughness, which we expect to be independent of time if the drop height is larger than the roughness.

The capillary force is

$$\sim \gamma (\theta_s^2 + \delta^2).$$

The gravity force is $\sim \rho g R^2 \theta_s^2$.

During the steps I, II, III, δ^2 is a constant.

Step I ("smooth regime")

The observed power law is well understood if in this step

$$\gamma \delta^2 \ll \gamma \theta_s^2, \ \rho g R^2 \theta_s^2 .$$

The drop spreads as if the surface was smooth. However, the crossover between capillarity and gravity is shifted towards short times.

Step II ("cap + foot regime")

Let us assume $\gamma \delta^2 \gg \gamma \theta_s^2$, $\rho g R^2 \theta_s^2$. For a drop spreading as a whole:

$$\gamma \delta^2 \sim \eta \theta^{-1} \frac{dR}{dt} = \text{cst}$$

$$R^3 \theta = \text{cst}$$

$$R(t) \sim t^{1/4} .$$

This power law is not observed. In fact, the cap goes on spreading with an unchanged power law and a foot develops. The unchanged power law for the cap

$$R_{\text{cap}} \sim t^{1/8} (\Omega_{\text{cap}} \gg \Omega_{\text{foot}})$$

can be understood if one considers that it is, in fact, spreading on a smooth surface. Another argument would be to treat this problem by analogy with the cap/precursor problem for smooth surfaces: the gravity term acts on the cap, and $\gamma \delta^2$ acts on the foot. (But this is just an ad hoc argument.)

Since the cap acts as a reservoir for the foot, the volume of the foot is not conserved. Its thickness e is of the order of the maximum roughness and does not change too much along its length ΔR.

So the friction is expected to be of the order of

$$\sim \frac{\eta}{e} \Delta R \frac{d \Delta R}{dt}$$

and balances the constant capillary force $\gamma \delta^2$. Finally

$$\Delta R(t) \sim \left(\frac{e}{\eta} t \right)^{1/2}$$

or

$$\Delta R(t) = (Dt)^{1/2}$$

equivalent to the Washburn equation [15] which governs imbibition phenomena [16].

We directly measured the length of the foot (in the case $\Omega_{\text{cap}} \gg \Omega_{\text{foot}}$) and plotted it versus $t^{1/2}$. The expected power law was observed. We repeated the experiment for several glass samples and several viscosities of the silicone oil. The results are summarized in Fig. 4 and 6. The relation

$$D \sim \eta^{-1}$$

is actually verified.

Finally, we expected $D \sim e \sim h$, where h is the maximum roughness of the glass. In Fig. 5, we have plotted $A = \eta D$ versus h. Obviously, our model is not perfect here, but the general trend $A \sim h$ is well observed.

It appears that step II is very rich, from a physical point of view. The apparent power law for the edge of the rim is, in fact, the superposition of two phenomena with well defined power laws.

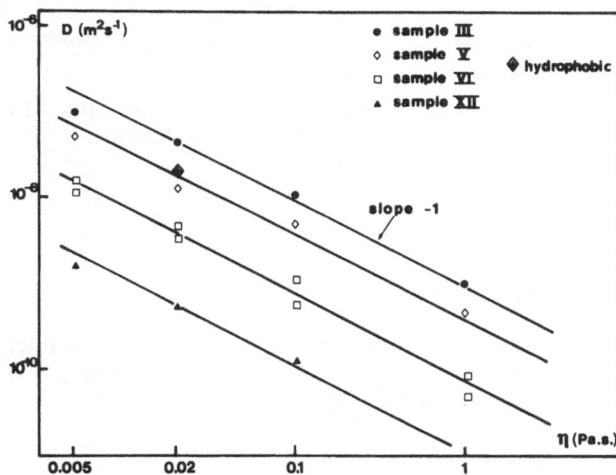

Fig. 4. Logarithmic plot of D versus η. D is defined by $\Delta R(t) = (Dt)^{1/2}$ and is expected to be A/η

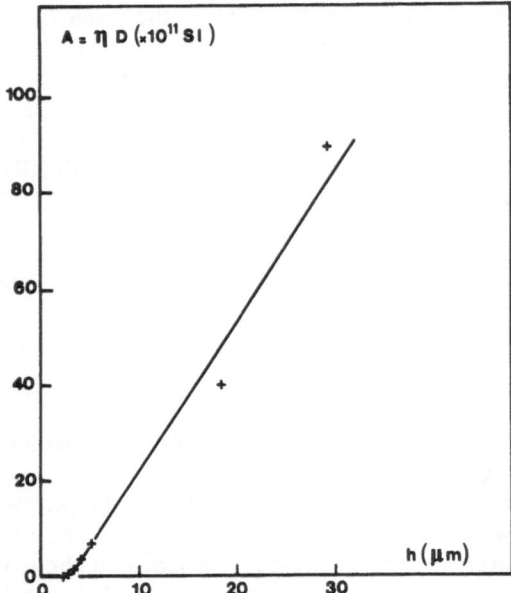

Fig. 5. Linear plot of $A = \eta D$ versus maximum roughness h

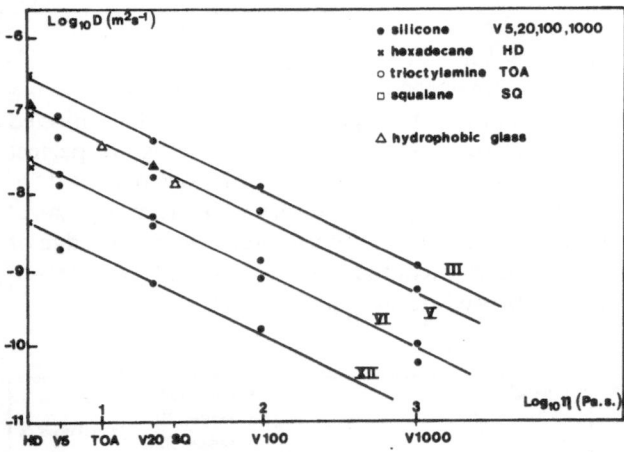

Fig. 6. Logarithmic plot of D versus η. D is defined by $\Delta R(t) = (Dt)^{1/2}$ and is expected to be A/η. Results for non-wetting liquids have been added

Step III ("flat drop")

The distinction between cap and foot is no longer visible. The drop spreads as a whole under the effect of the constant driving force $\gamma\delta^2$. So

$$R(t) \sim t^{1/4}$$

in good agreement with the experiment.

Step IV ("film regime")

The drop thickness is now much smaller than the glass roughness. We expect δ^2 to decrease with increasing time, but our simple model does not give quantitative information for this case. (Note that the experiment suggests $\delta^2 \sim R^{-2}$ which would give $x = 1/6 = 0.167$.)

IV. Comparison with other models

We will discuss now two more elaborated models available in the literature.

1. Model by Lenormand and Zarcone [17]

This model describes the filling of a porous medium connected to a reservoir. The medium is described as an ensemble of triangular ducts of maximum height H. This model adequately describes the feeding of the foot during step II, but the random character of the surface is not included.

The predicted length of the liquid in the ducts is found to be

$$l(t) = (Dt)^{1/2}$$

with $D \sim \dfrac{H}{\eta}$, as in our simple model.

2. Model by De Gennes [18]

For the moment we have no explanation for the power law obtained in step IV.

$$R(t) \sim t^x \quad x \sim 0.16$$

As the random character of the surface is important in this film regime, we tried to use the model of partial filling of fractal structures proposed by de Gennes.

It gives

$$R(t) \sim t^{0.2}.$$

The slowing down of the spreading is actually predicted. But the discrepancy with the experimental power law is significant. Clearly, a better description of the statistical properties of the roughness profile is needed.

V. Non-wetting situations

We will now briefly discuss situations in which a liquid that does not wet a smooth surface spreads on a rough one.

1. Experimental results

Non-wetting situations have been obtained with hexadecane on hydrophilic surfaces and with hexadecane, trioctylamin, squalane and triacetin on hydrophobic ones.

The contact angles θ_c of these liquids on smooth surfaces are finite (5° HD/hydrophilic glass; ~15° HD, trioctylamin, squalane/hydrophobic glass; 45° triacetin/hydrophobic glass).

Spreading is observed for all liquids, excepted for triacetin. During step II, the diffusion coefficient D was measured and found to be identical to its value for wetting situations (see Fig. 6) for all spreading liquids.

2. Discussion

The spreading condition is easily understood in the model of triangular ducts with angle $2a$. Spreading will occur if

$$\theta_c \leqslant \frac{\pi}{2} - a = \theta_0.$$

The roughness profile suggests $a \sim 60°-70°$ or $\theta_0 \sim 30°-40°$, larger than the measured contact angles, with the exception of triacetin. So the spreading condition holds well for these liquids.

With the same model, we would expect $\eta D = A$ to be strongly dependent on θ_c. Obviously this is not the case. The random character of the surface is probably the origin of this discrepancy. According to De Gennes [19], it might be independent of θ_c excepted for $\theta_0 \neq \theta_c$. θ_0 would be a percolation threshold for the spreading process and A expected to follow a well defined power law in this domain

$$A \sim (\theta_0 - \theta_c)^y \quad (y < 1)$$

$$A = 0 \quad \theta_c > \theta_0.$$

This theoretical prediction has to be compared with experimental results.

General conclusion

Despite the relatively long scientific history of the subject of wetting, many phenomena are not yet fully understood. As a complement to very elaborated calculations, it is useful to use simple models that emphasize the basic physical processes. De Gennes' approach [6] is very powerful from this point of view and allows a better understanding of the spreading phenomena. This is of primary importance for a wide range of practical applications.

Acknowledgements

It is a pleasure to acknowledge the workshop of the Ecole Supérieure d'Optique, where the rough surfaces were prepared and characterized; M. A. Guedeau for silane grafting and A. Bouillault who did some of the measurements. We are grateful to F. Brochard, P. G. de Gennes and M. Veyssié for helpful discussions, and R. J. Good, T. G. M. van de Ven and F. van Voorst Vader for providing us with bibliographic references.

References

1. Cahn J (1977) J Chem Phys 66:3667
2. Dussan EB (1979) V Ann Rev Fluid Mech 371
3. Neumann AW, Good RJ (1972) J Colloid Interface Sci 38:341; Eick JD, Good RJ, Neumann AW (1975) ibid. 53:235; Good RJ, Kotsidas ED (1978) ibid. 66:360
4. Lopez J, Miller CA, Ruckenstein E (1976) J Colloid Interface Sci 56:460
5. Teletzke GF, (1983) PhD Thesis, University of Minnesota, and references therein
6. De Gennes PG (1985) Rev Modern Physics 57:827; Joanny JF (1985) Thèse d'Etat, Paris
7. Tanner LH (1979) J Phys D 12, p 1473
8. Oliver JF, Mason SG (1973) The fundamental properties of paper related to its uses, Cambridge
9. Bayramli E, Van de Ven TGM, Mason SG (1981) Can J Chem 59:1954, 1962
10. Oliver JF, Mason SG (1977) J Colloid Interface Sci 60:480
11. Oliver JF, Mason SG (1980) J Mater Sci 15:431
12. Summ BD, Gorionov, Io V (1976) Chimiya (in Russian)
13. Cazabat, AM, Cohen Stuart MA, J Phys Chem (1986)
14. Cohen Stuart MA, Cazabat, AM, this issue, p 64
15. Washburn ED (1921) Phys Rev 17:374
16. Okagawa A, Mason SG (1977) In: Fibre-water interactions in paper-making, p 581, Univ. Press, Oxford
17. Lenormand R, Zarcone C (1984) 59th Ann Tech Cong and Exhibition – Soc of Petroleum Eng., Houston
18. De Gennes PG (1985) In: Adler D, Fritzsche H, Ovshinsky SR (eds) Physics of disordered materials. Plenum, New York
19. De Gennes PG, Private communication

Received September 22, 1986;
accepted November 28, 1986

Authors' address:

Cohen Stuart, M. A.
Agricultural University
Department for Physical and Colloid Chemistry
De Dreijen 6
NL-6703 BC Wageningen (Netherlands)

Progress in Colloid & Polymer Science   Progr Colloid & Polymer Sci 74:76–86 (1987)

Properties of a mean field fluid confined to a narrow slit

S. Sarman, J. C. Eriksson, R. Kjellander, and S. Ljunggren

Department of Physical Chemistry, The Royal Institute of Technology, Stockholm, Sweden

Abstract: The thermodynamic properties of an inhomogenous fluid confined to a narrow slit with attractive walls have been studied using a simple mean field theoretical approach. The pair-wise interaction assumed is a truncated Lennard-Jones potential with a repulsive hard core.

Density profiles of the fluid in the slit, the excess grand potential per unit surface area and the pressure tensor components p_N and p_T have been calculated. The lack of consistency among these variables, as well as the degree of agreement with the generalized Gibbs surface tension equation and with the Kelvin equation, is discussed. It is concluded that the (hard sphere) local density appproximation is a major source of the discrepancies noted.

Key words: Inhomogeneous Lennard-Jones fluid, pressure tensor, mean field approximation, density profiles, capillary condensation

Introduction

Statistical mechanical theories of capillary condensation in narrow pores and slits are of a fairly recent origin [1–4]. Most of these studies have been done by applying mean field theories or modified, non-local, mean field theories or, alternatively, by simulations. More accurate analytical methods such as the Percus-Yevick approximation [5] give rise to formidable computational problems [6] when the calculations are to be carried out at a constant chemical potential. In addition, it has been claimed by Evans et al. [7] that the Percus-Yevick closure is not applicable to the case of complete wetting at a solid-fluid interface.

It is generally assumed that, in spite of its obvious shortcomings, a mean field theoretical treatment can yield the main features of the phase transitions occurring in a pore as well as of the density profiles for the gas-like and liquid-like states involved.

In the main, our calculations are rather similar to those of some earlier authors [1–3]. Hence we shall concentrate on some more specific issues which have not been fully discussed before.

Model and computation methods

The system which we shall consider is a thin slit formed by two perfectly planar, parallel walls located at $z = 0$ and $z = h$. The fluid in the slit is open toward the surrounding bulk fluid where the chemical potential, μ, is determined. The slit is supposed to be large in the x- and y-directions. We may imagine its walls to be of circular shape. When there is a capillary condensate present in the slit there is also a quasi-cylindrical meniscus which is attached to the perimeter edges of the walls (Fig. 1).

For the pair potential, we have used a truncated Lennard-Jones potential with a hard core according to the expressions

$$u(r) = \begin{cases} 4\varepsilon(\sigma^{12}/r^{12} - \sigma^6/r^6) & \text{for} \quad r > \sigma \\ \infty & \text{for} \quad r < \sigma \end{cases} \tag{1}$$

The wall potential was assumed to be the sum of attractive contributions of the same form as in Eq. (1) between a molecule in the fluid and the particles of the wall which were supposed to be spherical and forming a hexagonal close-packed lattice, but with the potential averaged in the x and y directions. The complete expression for the attractive part of the wall potential has the following form

$$v(z) = -(\pi \varepsilon_w \sigma_w^6 \rho_w / 12)[1/z^3 + 1/(h-z)^3]. \tag{2}$$

Its repulsive part was represented by an impenetrable plane situated at the distance Δ in front of the outermost molecular plane of each wall.

In a similar way the wall-to-wall attractive potential was modelled as

$$v_{w/w}(h) = -A\sigma_w^2 \pi \varepsilon_{w/w} \sigma_{w/w}^6 / 3h^2 \qquad (3)$$

where h is the separation between the plates and A the surface area.

Our theoretical treatment is based on the following mean field model expression for the grand potential functional per unit surface area [4]

$$\Omega[\sigma] = \int dr_1 f_{hs}[\rho(r_1)] + 1/2 \int dr_1 dr_2 \rho(r_1)\rho(r_2)$$
$$\times u_{attr}(r_{12}) - \int dr_1 [\mu - v(r_1)]\rho(r_1) \qquad (4)$$

where $f_{hs}[\rho(r)]$ is the Helmholtz free energy density of a hard-sphere fluid with the (number) density $\rho(r_1)$. This means that the hard-sphere contribution to the free energy is evaluated in the local density approximation [8].

Minimization of the grand potential by requiring that

$$\delta\Omega/\delta\rho(r) = \delta F/\delta\rho(r) - \mu = 0 \qquad (5)$$

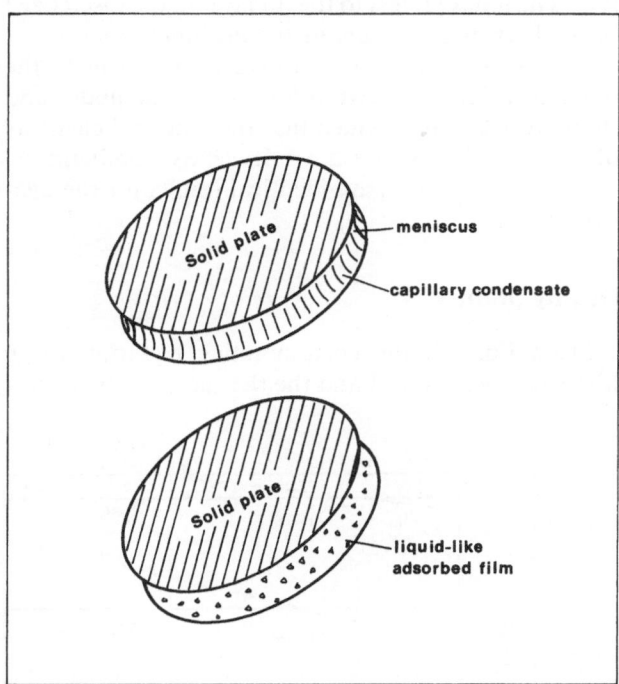

Fig. 1. Sketch of the slit system studied. At low vapour pressures (lower figure) the fluid in the slit is in a gas-like state with adsorbed liquid films on the walls, whereas at higher vapour pressures (upper figure) there is a capillary condensate in the slit which is limited by a curved meniscus

then yields the familiar expression

$$\mu = v(r_1) + \mu_{hs}[\rho(r_1)] + \int dr_2 \rho(r_2) u_{attr}(r_{12}) \qquad (6)$$

for the (constant) chemical potential μ where the integral of u_{attr} refers to the volume outside the hard core. The expression for the hard-sphere part, μ_{hs}, may be partitioned into two terms

$$\mu_{hs}[(r)] = kT \ln [\rho(r)\Lambda^3] + \mu_{hs}^{ex}[\rho(r)] \qquad (7)$$

where the first term is the ideal contribution, Λ being, as usual, de Broglie's thermal wavelength. For the second (excess) term we have adopted an expression derived from the Percus-Yevick equation for a homogeneous fluid using the virial theorem [5].

$$\mu_{hs}^{ex}/kT = (5\eta^3 - 13\eta^2 + 14\eta)/2(1-\eta)^3 - \ln(1-\eta) \qquad (8)$$

where $\eta = \pi\rho\sigma^3/6$ denotes the volume fraction.

Performing the integration in Eq. (6) for the case $v = 0$ and at constant ρ, the following expression for the chemical potential in the homogeneous phase emerges

$$\mu(\eta)/kT = \ln\eta - \ln(1-\eta) + (5\eta^3 - 13\eta^2 + 14\eta)/$$
$$2(1-\eta)^3 - 64\varepsilon\eta/3kT . \qquad (9)$$

Subtracting from Eq. (6) the corresponding equation for the bulk phase of density ρ_0 with the same chemical potential μ, we obtain

$$-kT \ln[\rho(r_1)/\rho_0] - v(r_1) + \int dr_2 u_{attr}(r_{12})[\rho_0 - \rho(r_2)]$$
$$-\mu_{hs}^{ex}[\rho(r_1)] + \mu_{hs}^{ex}(\rho_0) = 0 \qquad (10)$$

or, in terms of the volume fraction (also taking into account the fact that the density varies only in the z direction),

$$-kT \ln[\eta(z_1)/\eta_0] - v(z_1) + \int_0^h dz[\eta_0 - \eta(z)] U(z-z_1)$$
$$+ \int_{-\infty}^0 dz\eta_0 U(z-z_1) + \int_h^\infty dz\eta_0 U(z-z_1) + \mu_{hs}^{ex}(\eta_0)$$
$$-\mu_{hs}^{ex}[\eta(z_1)] = 0 \qquad (11)$$

where

$$U(z) = \begin{cases} -48.000/(z/\sigma)^4 - 2457.61/(z/\sigma)^{10} & \text{for } z > \sigma \\ -3.6000 & \text{for } z < \sigma \end{cases}$$
$$(12)$$

The partitioning of the integrals in Eq. (11) has the effect of stabilizing the equations numerically.

The grand potential per unit surface area may be written

$$\Omega = -p_e h + \gamma \tag{13}$$

where p_e is the external pressure in the homogeneous bulk phase and γ is the interfacial free energy or a generalized "surface tension" of the fluid in the slit (fluid film tension), not to be confused with the ordinary *sl, sg* or *lg* surface tensions. We thus obtain

$$\gamma = \Omega + p_e h = \int\limits_0^h dz \{\rho(z) \mu_{hs} [\rho(z)] - p_{hs} [\rho(z)]\}$$
$$+ 1/2 \int\limits_0^h \int\limits_0^h dz\, dz'\, U(z-z') \rho(z) \rho(z')$$
$$- \int\limits_0^h dz\, \rho(z) [\mu - v(z)] + p_e h \ . \tag{14}$$

Elimination of μ between Eq. (14) and the z-dependent form of Eq. (6) finally yields an attractively simple expression for γ, viz:

$$\gamma = \int\limits_0^h dz \{p_e - p_{hs} [\rho(z)]\} - 1/2 \int\limits_0^h \int\limits_0^h dz\, dz'\, U(z-z') \rho(z) \rho(z') \tag{15}$$

involving a hard-sphere term and an internal fluid interaction term. $p_{hs}(\rho)$ is the pressure of a hard-sphere fluid of density ρ. If there are two or more density profiles satisfying Eq. (6) or (11) at the same chemical potential, it is well known that the stable profile is the one with the lowest grand potential Ω. Since p_e is also the same when μ is the same, it follows from Eq. (13) that we may equally well use the value of γ as a stability criterion. The mechanical interpretation of this is that the slit-fluid with the lower γ-value will "squeeze out" the slit-fluid with the higher γ-value.

Improved numerical accuracy resulted from the use of dimensionless variables, i.e. μ/kT for the chemical potential, $p\pi\sigma^3/6kT$ for the pressure, $\gamma\pi\sigma^2/3kT$ for the interfacial free energy, the volume fraction $\eta = \pi\rho\sigma^3/6$ for the density and $\beta\varepsilon = \varepsilon/kT$ for the inverse temperature.

The integrals in Eq. (11) were calculated in two different ways: by the NAG (Numerical Algorithm's Group) Fortran program library routine DO1GAF and by the Romberg method of iterated Richardson extrapolation. The Richardson extrapolation was repeated four times. The former method is the more accurate one but is more time-consuming.

The solution of the integral Eq. (11) was also performed in two different ways. The first method was based on the NAG routine CO5PCF which has the advantage of being robust and fast. On the other hand, it requires plenty of storage space and does not permit the use of the most accurate integration method.

The second method was based on the following iteration scheme:

$$\eta_{n+1}(z_i) = \eta_n(z_i) - \eta_n(z_i) [1 - \eta_n(z_i)]^4 G[\eta_n(z_i)]/$$
$$[1 + 2\eta_n(z_i)]^2 \tag{16}$$

where the function G denotes the left hand side of Eq. (11). The index i refers to the ith layer in the discretized integral whereas n refers to the nth iteration.

This iteration method is roughly equivalent to the one-dimensional Newton-Raphson method and should work well provided that the diagonal elements of the Jacobian matrix are sufficiently dominant.

The two methods gave identical results for the density profile.

Density profiles

From Eq. (9), the corresponding equation for p [derived from Eq. (9)] and the thermodynamic identi-

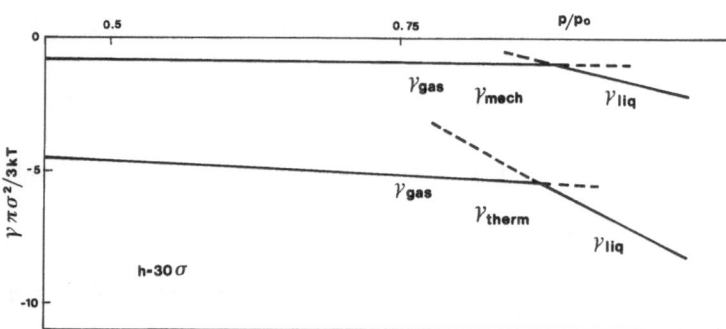

Fig. 2a. γ_{mech} and γ_{therm} plotted as functions of p/p_0 at constant slit width $h = 30\sigma$

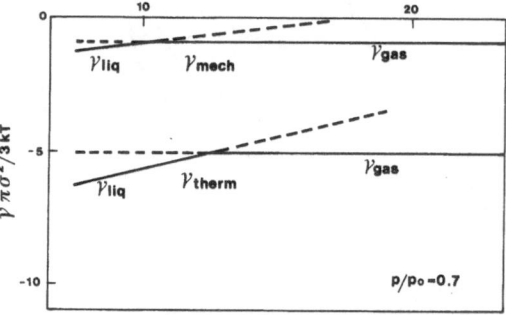

Fig. 2b. γ_{mech} and γ_{therm} plotted as functions of the slit width h at $p/p_0 = 0.7$

ty $\partial p/\partial\rho = \rho\,\partial\mu/\partial\rho$ it can be shown that to within less than one part in a thousand, $\varepsilon = kT_c$, T_c denoting the critical temperature of the bulk fluid. Since the calculations are carried out in terms of dimensionless variables, the only remaining variable parameters are then $\varepsilon_w/\varepsilon$, Δ, T/T_c, p/p_0 and h/σ where p_0 is the saturation vapour pressure.

In the calculations referred to in this paper Δ was set equal to 1.7 molecular diameters, $\varepsilon_w/\varepsilon$ equal to 2 and T/T_c equal to 0.625. The relative vapour pressure p/p_0 was varied from case to case. The same thing holds for the plate-to-plate distance, h.

No prewetting was observed under the conditions chosen for these calculations. Figs. 2a and b show the values of γ calculated under various conditions. The

designation γ_{therm} refers to the excess free energy γ as calculated by means of Eq. (15). The plot illustrates how γ can be used to judge which of the two possible states ("gas" or "liquid") is thermodynamically stable. Thus, the gas-like state is stable at low p/p_0 and large separations, whereas the liquid-like state is stable at high p/p_0 and small separations. The mechanically defined film tension γ_{mech}, plotted in the same diagram, is discussed in a subsequent paragraph.

Figures 3a−c show some typical density profiles obtained under various conditions. These curves are similar to those derived earlier by other authors [2−3] and will not be further discussed here. It may be of some interest to note, however, that there is a hint of a knee in the curve of Fig. 3b. For thicker adsorption films this feature is even more pronounced. Hence, the adsorbed fluid film has an advantageous inner part, next to the wall, and a disadvantageous outer part. As might be anticipated for a mean field model, no real density oscillations are seen.

We can define the slit adsorption in the following way

$$\Gamma = \int_0^h dz\,[\rho(T,\mu,z) - \rho_0(T,\mu)]\;. \qquad (17)$$

Figure 4 shows Γ as a function of p/p_0 at $h = 12\sigma$. For large values of h, the asorption vs. p/p_0 curve resembles the BET isotherm until condensation occurs in the slit. The agreement with the BET shape becomes increasingly more accurate at low p/p_0. This shape reflects, of course, the strong repulsions which rapidly develop in the liquid adsorbate close to the

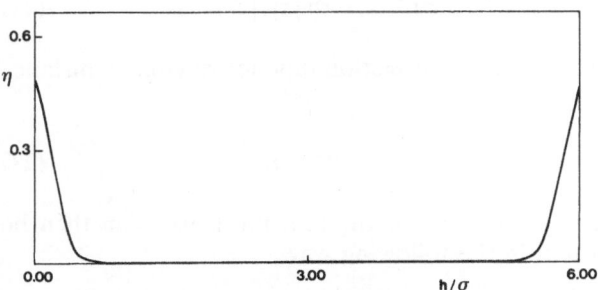

Fig. 3a. Density profile for a gas-like state of the slit fluid at $p/p_0 = 0.025$ and $h = 6\sigma$

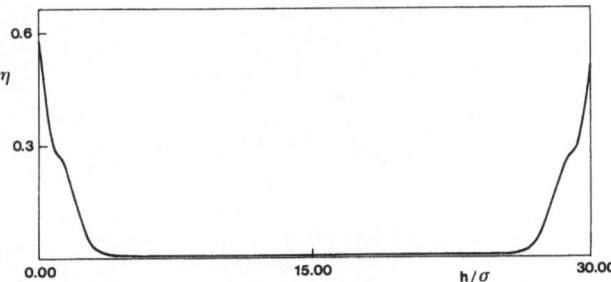

Fig. 3b. Density profile for a gas-like state of the slit fluid at $p/p_0 = 0.7$ and $h = 30\sigma$

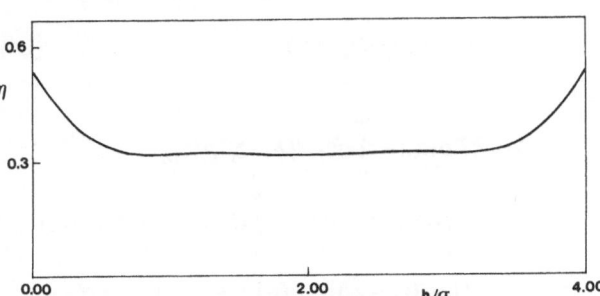

Fig. 3c. Density profile for a capillary-condensed slit fluid at $p/p_0 = 0.3$ and $h = 4\sigma$

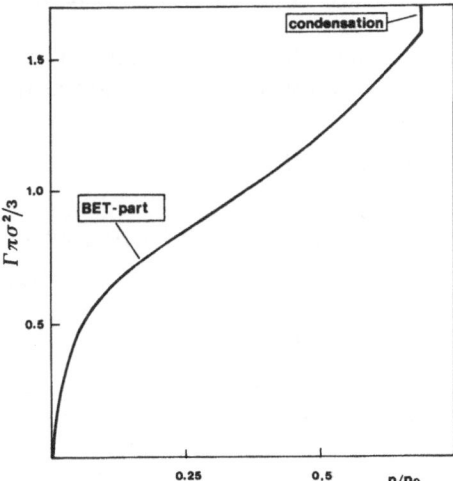

Fig. 4. Adsorption isotherm obtained before capillary condensation for a slit with $h = 12\sigma$

walls and the much weaker wall potential felt by the ad-molecules at some distance away from the walls.

The pressure tensor and the film tension

There is an inherent ambiguity in the definition of the pressure tensor, basically because the inter-molecular forces act between molecules and not upon mathematical surfaces. This means that the divergence of the pressure tensor is always well-defined but that the pressure tensor itself is not. Hence there are several different definitions of the pressure tensor, the most well-known of which are associated with the names of Kirkwood and Buff [9], Harasima [10] and Irving and Kirkwood [11]. The general problem has been treated by Henderson and Schofield [12].

This ambiguity is effectively of no consequence when considering bodies of large dimensions with short-range intermolecular forces. But, in the case of a very narrow slit, the ambiguity of the mechanically defined pressure tensor requires careful consideration. It is easy to show, however, that in the geometry of a slit with plane-parallel walls the normal pressure component, $P_N = p_{zz}$, is invariant to the convention chosen for its definition, i.e. Harasima's definition and that of Irving and Kirkwood, etc., all' yield the same value of p_N. Similarly, the mechanical definition of γ, i.e..

$$\gamma_{mech} = \int_0^h dz \, [p_e - p_T(z)] \tag{18}$$

where p_T is the tangential component of the pressure tensor, is also convention-invariant. The pressure tensor is also rather meaningful at the centre of a fairly wide slit, where ρ is nearly constant over a range of z-values.

According to Harasima [10], the tangential and normal components of the pressure tensor are given by

$$p_T(z_1) = kT\rho(z_1) - 1/4 \iiint_V d^3r_{12} u'(r_{12})$$
$$\times \frac{x_{12}^2 + y_{12}^2}{r_{12}} g(s_{12}, z_1, z_1 + z_{12})\rho(z_1)\rho(z_1 + z_{12}) \tag{19}$$

and

$$p_N(z_0) = kT\rho(z_0) - \int_0^{z_0} dz_1 v'(h - z_1)\rho(z_1)$$
$$- \int_{z_0}^h dz_1 v'(z_1)\rho(z_1) - \int_{z_0}^h dz_1 \rho(z_1)$$
$$\times \int_0^{z_0} dz_2 \rho(z_2) \int ds_{12} g(s_{12}, z_1, z_2)(\partial u(r_{12})/\partial z_1) \tag{20}$$

where we have excluded the direct wall/wall attraction and where s_{12} is the vector with the components x_{12} and y_{12}.

In order to evaluate these expressions in accordance with the general philosophy of the mean field theory, the product of the derivative of u and the pair correlation function, g, is subdivided into two parts, one for the attractive part where g is set equal to one, and one for the hard core part where g is taken from the Percus-Yevick equation for hard spheres. It is then easy to show that

$$g_{hs}(r)(\partial u_{hs}(r)/\partial r) = -kT g_{hs}(\sigma)\delta(r - \sigma) \tag{21}$$

and

$$g_{hs}(r_{12})(\partial u_{hs}(r_{12})/\partial z_1)$$
$$= -kT g_{hs}(\sigma)\delta(r_{12} - \sigma) |z_{12}|/r_{12} \tag{22}$$

where the pair correlation function at contact distance is given by

$$g_{hs}(\sigma) = (4 - 2\eta + \eta^2)/4(1 - \eta)^3 . \tag{23}$$

The components of the pressure tensor can then be written in the following way

$$p_T(z_1) = kT\rho(z_1) - \tfrac{1}{2}\pi\rho(z_1)\int_0^h dz_2\rho(z_2) \int_{\max(|z_{12}|, \sigma)}^{\infty} dr_{12}$$
$$\times u'_{attr}(r_{12})(r_{12}^2 - z_{12}^2)$$
$$+ \tfrac{1}{2}kT\rho(z_1) \int_{\max(-\sigma, -z_1)}^{\min(\sigma, h - z_1)} dz_{12}$$
$$\times g_{hs}\left[\sigma, \rho\left(\frac{z_1 + z_2}{2}\right)\right]\rho(z_2)(\sigma^2 - z_{12}^2)$$
$$= kT\rho(z_1) + \tfrac{1}{2}kT\rho(z_1) \int_{\max(-\sigma, -z_1)}^{\min(\sigma, h - z_1)} dz_{12}$$
$$\times \rho(z_2)(\sigma^2 - z_{12}^2) g_{hs}\left[\sigma, \rho\left(\frac{z_1 + z_2}{2}\right)\right]$$
$$+ \tfrac{1}{2}\rho(z_1)\int_0^h dz_2 \rho(z_2) U(z_1 - z_2) \tag{24}$$

and

$$p_N(z_0) = kT\rho(z_0) - \int_0^{z_0} dz_1 v'(h - z_1)\rho(z_1)$$
$$- \int_{z_0}^h dz_1 v'(z_1)\rho(z_1) - \int_{z_0}^h dz_1 \rho(z_1) \int_0^{z_0} dz_2 \rho(z_2)$$
$$\times [2\pi(z_1 - z_2)u_{attr}(\sigma) + 2\pi kT |z_1 - z_2|$$
$$\times g_{hs}(\sigma)\theta(z_1 - z_2)] \tag{25}$$

where $\theta(z)$ is equal to 0 for $|z| > \sigma$ and equal to 1 for $|z| < \sigma$.

It is reassuring to note that in the homogeneous bulk phase both p_T and p_N, as calculated from Eqs. (24) and (25), become equal to the bulk pressure derived from Eq. (9) by using the thermodynamic relation $\partial p/\partial \rho = \rho(\partial \mu/\partial \rho)$:

$$\pi \sigma^3 p_{\text{bulk}}/6kT = (\eta + \eta^2 + \eta^3)/(1-\eta)^3 - 32\varepsilon\eta^2/3kT \ . \tag{26}$$

Thus in the bulk phase we have full consistency.

Another way of calculating p_N is to use thermodynamic relation [13, 14]

$$\pi = p_N - p_e = -(\partial \gamma/\partial h)_{T,\mu} \tag{27}$$

applied to the expression of γ in Eq. (15). We shall use the designation $(p_N)_{\text{therm}}$ for the value of p_N derived from Eq. (27).

Comparison between Eqs. (15) and (18) also shows that, somewhat vaguely, we may define a "thermodynamic" value for the tangential pressure component as:

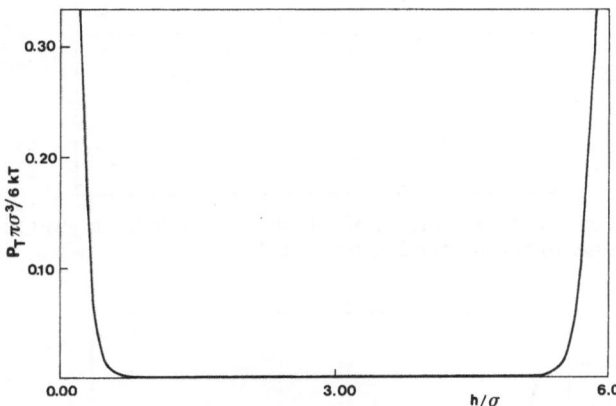

Fig. 5a. P_T-profile for a gas-like state at $p/p_0 = 0.025$ and $h = 6\sigma$

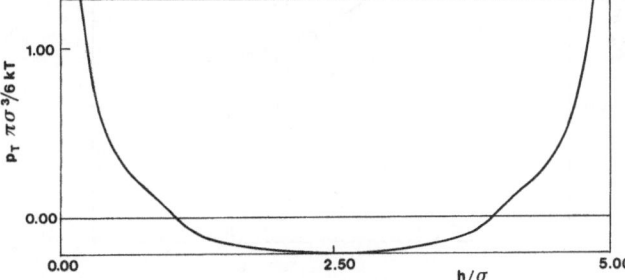

Fig. 5b. P_T-profile for a liquid-like state at $p/p_0 = 0.3$ and $h = 5\sigma$

$$p_T(z) = p_{hs}[\rho(z)] - 1/2\,\rho(z)\int_0^h dz'\rho(z')\,U(z-z') \ . \tag{28}$$

Comparing with Eq. (24), we note that the attractive part of p_T is the same in the two expressions, the difference being that a local approximation for p_{hs} is used in Eq. (28) while the corresponding expression in Eq. (24) is basically non-local. Actually, Eq. (28) illustrates the main shortcomings of the mean field approach rather clearly. The local density approximation places the whole anisotropy of the pressure tensor in the attractive part of p_T.

The difference between p_N as obtained from Eq. (27) and as calculated for the middle plane in the slit using Eq. (25) is not very pronounced. For wide slits $(h > 8\sigma)$ it is less than 10%, whereas for narrow slits $(h < 8\sigma)$ discrepancies of a greater magnitude are noted.

Calculations based upon the p_T expression, Eq. (24), show that there are two different types of tangential pressure profiles, depending on whether one is dealing with a gas-like or liquid-like solution. Figure 5 shows the pressure profiles for a gas-like solution at a low pressure, i.e. at a pressure equal to less than 10% of the saturation pressure, p_0. The tangential pressure p_T may reach a few thousand atmospheres close to the walls but falls off to the pressure of the homogeneous bulk phase at the centre of the slit. When the saturation pressure is approached, a region with a negative tangential pressure appears at a distance of a few molecular radii from the wall. This region coincides with the outermost part of the adsorption layer, and the negative tangential pressure can then be associated with the "surface tension" of the adsorbed liquid film, in contact with the gas in the middle of the slit.

Figure 5b shows the p_T-profile of a liquid-like solution. The pressure close to the walls amounts to some thousand atmospheres and falls off towards the centre of the slit. At the very centre p_T is negative when the liquid actually constitutes (or closely resembles) a capillary condensate. This negative pressure may vary between zero and a few hundred atmospheres, depending upon the ratio p/p_0. Moreover, for capillary-condensate states, the normal and tangential pressures, p_N and p_T, are at least approximately equal in the midplane at $z = h/2$ (cf. Fig. 6). For a slit with a width of $h = 5\sigma$ the difference between p_N and p_T amounts to about 10% and decreases when the width of the slit is increased. Thus the degree of anisotropy of the pressure tensor is generally rather insignificant in the mid-plane of a slit, unless its width is very narrow.

The normal pressure, p_N, is, of course, independent of z. The variation of p_N with the slit width, h, is illustrated in Fig. 7a for a gas-like solution. The diagram shows how the normal pressure converges toward the isotropic pressure of the homogeneous bulk phase when the distance between the plates increases.

When the plates approach each other, p_N changes sign and becomes increasingly more negative. The reason for this is that at large distances between the plates there is a nearly homogeneous gas phase between the plates which is only weakly affected by the wall potentials and by the adsorption layers at the wall. The properties of this gas phase should, therefore, not deviate to any great extent from those of a homogeneous gas phase. A shorter distances between the plates, the adsorption layers on the opposite walls approach one another and the opposite wall. They will then interact, giving rise to an attractive normal pressure.

Figure 7b shows the h-dependence of p_N for a liquid-like solution. At large distances between the plates, p_N approaches the pressure of the corresponding homogeneous phase just as in the case of the gas-like solution. When the distance between the plates decreases, the normal pressure will rise since the lowering of the potential energy, due to the negative wall potential, must be compensated by a corresponding pressure rise in order to keep the chemical potential constant. This reasoning is based upon the observation that $p_N \approx p_T$ in the mid-plane of the slit (cf. Fig. 6). In other words, the attractive wall potential tends to pull molecules from the surrounding bulk phase into the slit, thereby raising the pressure level.

Figure 7c shows how p_N depends on the chemical potential at a constant distance between the plates and at constant temperature. For convenience, p/p_0 has been used as an independent variable rather than the chemical potential itself.

At large distances between the plates, the normal pressure increases linearly with the density of the homogeneous phase until the condensation point is reached. This is understandable, since the normal pressure is rougly equal to the pressure of the homogeneous phase at large h and the latter pressure is proportional to the density at low densities. When the distance between the plates is decreased, the normal pressure will rise less rapidly with the density in the bulk phase. At a certain distance between the plates, it becomes practically equal to zero, independently of p/p_0; upon further decrease of the distance between the plates, the normal pressure will decrease with p/p_0. This is due to the same reason that gives rise to a negative pressure in the gas phase

when the plates approach each other. At the condensation point a liquid is formed with a negative p_N with an absolute value tens of times larger than the

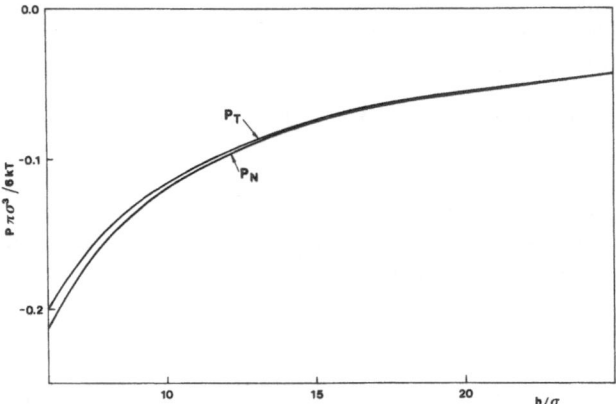

Fig. 6. P_T and P_N at the centre of the slit fluid in the liquid state at the condensation point

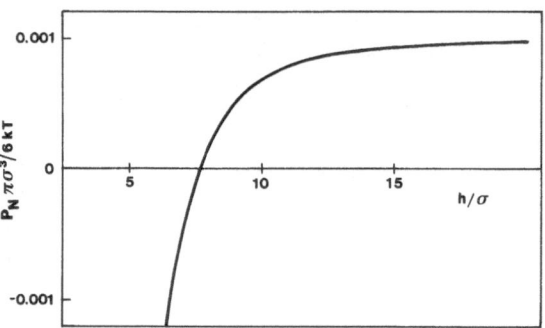

Fig. 7a. P_N as a function of the slit width for a gas-like state of the slit fluid at $p/p_0 = 0.2$

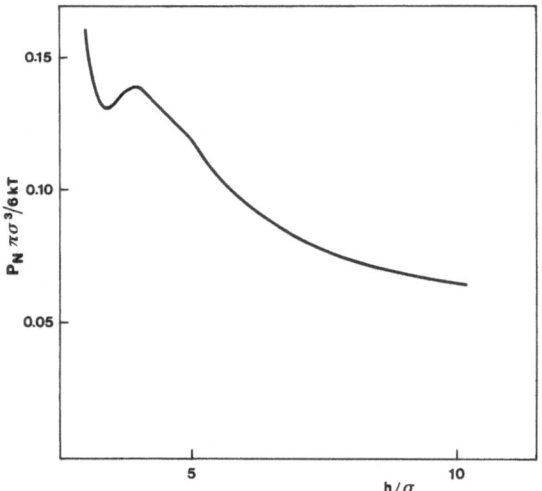

Fig. 7b. P_N as a function of the slit width for a liquid-like state of the slit fluid at $p/p_0 = 10$

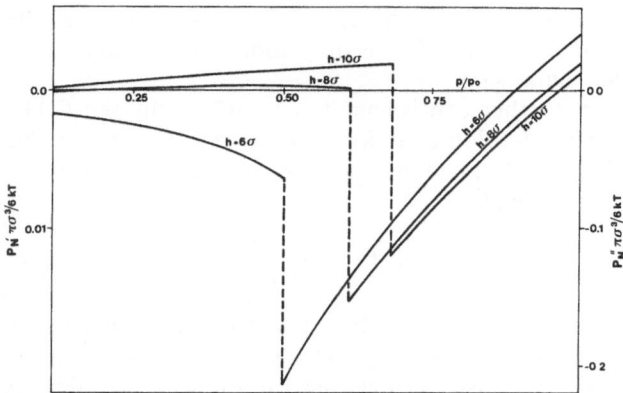

Fig. 7c. P_N as a function of the p/p_0 for various slit widths. The scale to the left, p'_N, refers to the gas-like state and the scale to the right, p''_N, to the liquid-like state

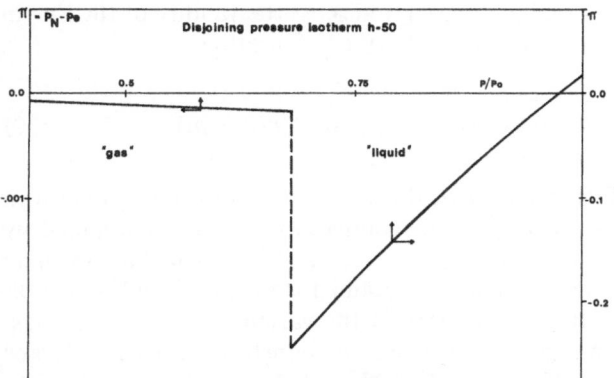

Fig. 7d. Disjoining pressure $\pi = p_N - p_e$ as a function of p/p_0 for $h = 50\sigma$

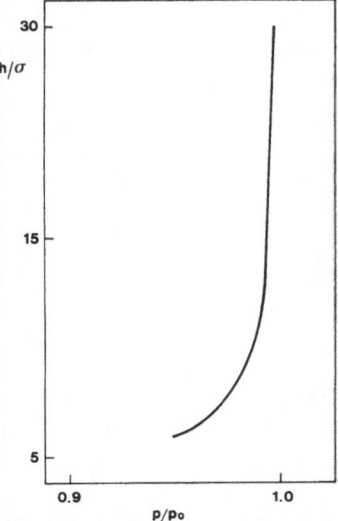

Fig. 8. Equilibrium thickness, h_{eq}, of a capillary condensate as a function of p/p_0. Note that the wall to wall interactions are neglected

(positive) normal pressure of a gas with the same chemical potential. However, the normal pressure increases rapidly with p/p_0 and becomes positive before the condensation point of the homogeneous bulk phase is reached. The shorter the distance between the plates, the more negative the normal pressure is at the condensation point in the slit. However, the process is qualitatively independent of the distance between the plates.

A further look at Fig. 7d shows that the disjoining pressure, $\pi = p_N - p_e$, at the condensation point in the slit is negative, but increases rapidly and becomes zero before $p/p_0 = 1$ is reached. At this point, no net force is acting between the plates except their mutual wall/wall attraction, which is small at large distances. It is also apparent from Fig. 7d that this point represents a stable equilibrium position. If there is no external force acting on the plates they will, therefore, adjust themselves so as to attain this equilibrium state. The equilibrium distance, h_{eq} for which $\pi = 0$, varies with the density of the bulk phase, i.e. with the chemical potential (cf. Fig. 8). This phenomenon bears a certain resemblance to the swelling of lamellar liquid crystals [15] although in that case different kinds of forces are operative, such as hydration forces and hydrophobic forces. It is worth noting that h_{eq} increases quite rapidly with p/p_0 when $p/p_0 = 1$ is approached. Hence, a behaviour of this kind need not necessarily be associated with interactions that are intrinsically long-range. What we are dealing with here can be viewed as a balance between the pressure-rising effect of the wall potential and the (negative) capillary pressure due to the meniscus, of quasi-cylindrical shape, at the perimeter of the slit.

Consistency of the model and validity of the macroscopic Laplace, Kelvin and Gibbs equations

As Figs. 2a and 2b show, there is a large discrepancy between the thermodynamic γ calculated from Eq. (15), γ_{therm}, and Eq. (24) and the mechanically defined γ calculated from Eq. (18), γ_{mech}. Since γ is invariant to the convention used to define p_T, this discrepancy cannot be due to the lack of localization of the pressure tensor, but must be due to an intrinsic inconsistency of the mean field method itself. The main part of this inconsistency evidently arises in the immediate vicinity of the walls, where the density varies rapidly with z, since the formula for p_T, Eq. (24), is known to give results that are nearly consistent with Eq. (26) in the almost homogeneous region at the centre of the slit. In spite of this inconsistency, the positions of the transition points between the gas- and liquid-like solutions are nearly the same for γ_{therm} and γ_{mech}.

Another consistency test is based on Eq. (27) with γ calculated either as γ_{therm} or as γ_{mech}. In the case of a gas-like solution this equation is fulfilled for γ_{therm} as well as γ_{mech}. The reason is apparently that the value of γ in the gas-like solution is practically independent of h, as long as the adsorption layers do not overlap and the normal pressure is almost equal to the external pressure in the bulk phase, p_e. In the case of the liquid-like solution, Eq. (27) is satisfied by γ_{therm}

but definitely not by γ_{mech} as in the latter case the right side of Eq. (27) is only about half as large as the left hand side.

A further consistency test is offered by the Gibbs surface tension equation, generalized so as to involve the plate-to-plate distance h as an additional independent variable:

$$\partial\gamma(T, \mu, h)/\partial\mu = -\Gamma \qquad (29)$$

where Γ is defined by Eq. (17). This equation is found to be satisfied by γ_{therm}. This is only natural, however, since it is easy to show that for any solution which is based on minimizing the grand potential according to Eq. (5), γ_{therm}, defined by Eq. (13), will automatically satisfy Eq. (29). On the other hand, Eq. (29) is definitely not satisfied by γ_{mech}.

Finally, we need to discuss the validity of the Kelvin equation for capillary condensation:

$$kT \ln (p/p_0) = -2\gamma_{\text{lg}} \cos \theta/h(\rho_1 - \rho_g) . \qquad (30)$$

Figure 9a shows the step-wise adsorption isotherm in the vicinity of the condensation point determined by the condition $\gamma_{\text{gas}} = \gamma_{\text{liquid}}$. From such data we may construct phase diagrams for the gas-liquid transition in the slit in terms of the variables p/p_0 and h, Figs. 9b and 9c, as has been demonstrated earlier by Evans and Tarazona [1]. Note that we do not need to separately know the liquid-gas surface tension γ_{lg} and the contact angle θ in this context, assuming that the macroscopic Eq. (30) yields the correct limiting

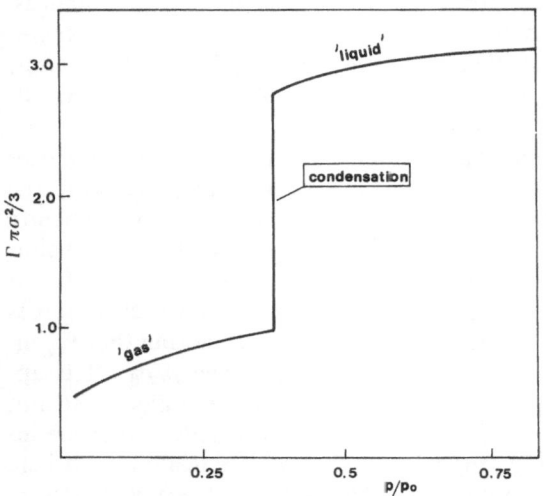

Fig. 9a. Adsorption isotherm for a 5σ wide slit, showing a first order, stepwise condensation from "gas" to "liquid"

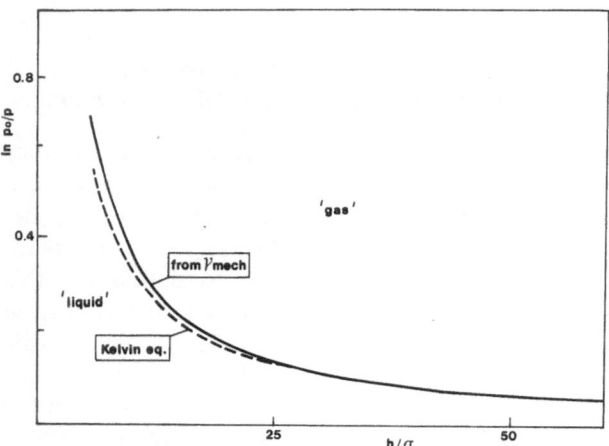

Fig. 9b. Coexistence line for "gas" and "liquid" in the slit using γ_{mech}, compared with the Kelvin-equation-based prediction ($---$). For large h the latter implies that $\gamma_{\text{lg}} \cos \theta \times \pi \sigma^2/3kT = 1.26$

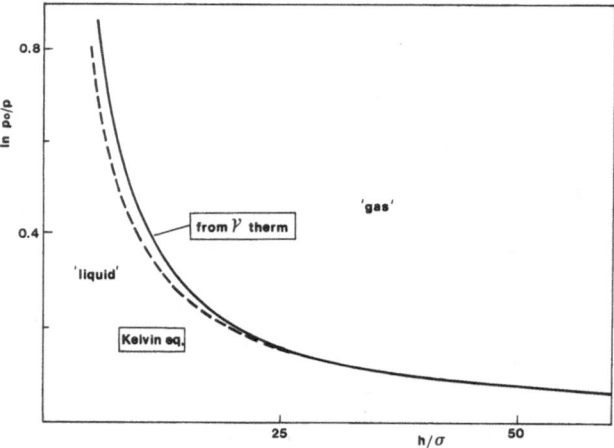

Fig. 9c. Coexistence line for "gas" and "liquid" in the slit using γ_{therm}, compared with the Kelvin-equation-based prediction ($---$). For large h the latter implies that $\gamma_{\text{lg}} \cos \theta \times \pi \sigma^2/3kT = 1.47$

behaviour at large *h*. This means that the $\ln (p/p_0)$ versus *h* function obtained from Eq. (30) has been assumed to coincide at large *h* with the corresponding functions generated by means of the condition $\gamma_{gas} = \gamma_{liquid}$. Furthermore, we anticipate here that the true equilibrium contact angle can always be established at the rim edges of the slit walls.

As appears from the Figs. 9b, c, the agreement between the Kelvin equation and our mean field calculations is quite good, as tested in this manner at distances exceeding 15 to 25σ. At smaller slit widths the deviation is more pronounced. However, since the mean field calculations do not correctly reproduce the density oscillations near the wall, it is unlikely that any more definite conclusions regarding the range of validity of the Kelvin equation may be drawn from these calculations. Still it might be argued, referring to the calculation results presented here, that the ordinary Laplace equation (on which the derivation of the Kelvin equation is based) should provide a reasonably good estimate of the pressure tensor components p_T and p_N in the centre of the slit (where the density oscillations have their smallest amplitudes [13]) insofar as these components are not very different in magnitude. Under such conditions the Kelvin equation is expected to be approximately valid as well, even for rather narrow slits ($h > 5\sigma$). The experimental investigations carried out by Israelachvili and Fischer [16] also substantiate this provisional conclusion.

Concluding remarks

The inconsistencies discussed in this paper must evidently have their origin in the two main approximations of the mean field theory, viz: the use of a local density approximation [8] for the hard sphere contribution to the chemical potential, and the artifice of setting the pair correlation equal to unity in the attractive contribution to the free energy. The first approximation is probably the most serious, since the second approximation has already been partly compensated for by a suitable choice of parameters.

Variants of the mean field treatment where the density, ρ, in the expression for the hard sphere contribution to the chemical potential, $\mu_{hs}(\rho)$, is replaced by the mean value of the density in a small spherical volume surrounding the point in question, have been suggested, for example by Nordholm [17] and by Evans and Tarazona [18] and probably constitute a considerable improvement on the original mean field theory. In contrast to the latter, these improved methods are even capable of reproducing the density

oscillations near the plate surfaces in a seemingly correct fashion. However, in order to test the validity of these approaches and in order to obtain complete consistency there is still an urgent need for statistical mechanical treatments using more sophisticated methods.

One major stumbling block in connection with all such treatments is the necessity of carrying out the calculations at a constant chemical potential which entails a difficult integration over a coupling parameter. As in the case of the Percus-Yevick equation, this may lead to an infinite cluster expansion [6]. Another unanswered question is the problem of the validity of different theoretical approaches when applied to adsorption or capillary condensation phenomena [7].

Apart from all these methodological difficulties we still have the basic problem of whether a theoretical calculation for a perfect slit-shaped or cylindrical pore with perfectly planar walls bears any relation to capillary condensation in a real porous medium with irregularly shaped pores. It may well be that capillary condensation in a real pore includes important features which are more closely related to what happens in a molecular sieve and that the idealization of a perfect pore with atomically smooth walls is inadequate. Furthermore, the detailed conditions at the pore openings are likely to be of great significance for the mechanical equilibrium of a capillary condensate and these are undoubtedly very difficult to assess.

References

1. Evans R, Tarazona P (1984) Phys Rev Lett 52:557
2. Evans R, Marconi UMB, Tarazona P (1986) J Chem Soc, Faraday Trans 2, 82:1763
3. Peterson BK, Walton JPRB, Gubbins KE (1986) J Chem Soc, Faraday Trans 2, 82:1789
4. Rowlinson JS, Widom B (1984) Molecular theory of Capillarity. Oxford Univ Press
5. Wertheim M (1964) J Math Phys 5:643
6. Morita T, Hiroike K (1960) Progr Theor Phys 23:1003
7. Evans R, Tarazona P, Marconi UMB (1983) Mol Phys 50:993
8. Sullivan DE (1981) J Chem Phys 74:2604
9. Kirkwood JG, Buff FP (1949) J Chem Phys 17:338
10. Harasima A (1958) Adv Chem Phys 1:203
11. Irving JH, Kirkwood JG (1950) J Chem Phys 18:817
12. Schofield P, Henderson JR (1982) Proc R Soc Lond A 379:231
13. Magda JJ, Tirrell M, Davis T (1985) J Chem Phys 83:1888
14. Ash SG, Everett DH, Radke CJ (1973) J Chem Soc, Faraday Trans 2, 69:1256
15. Eriksson JC, Ljunggren S (1986) Acta Chem Scand A 40:261

16. Israelachvili JN, Fischer LP (1981) Colloids Surf 3:303
17. Freasier BC, Nordholm S (1985) Mol Phys 54:33
18. Tarazona P, Evans R (1984) Mol Phys 52:847

Received October 28, 1986;
accepted November 28, 1986

Authors' address:

Sten Sarman
Department of Physical Chemistry
The Royal Institute of Technology
S-10044 Stockholm, Sweden

Solubility and phase behaviour of ferric dodecyl benzene sulphonate in aqueous solutions

Đ. Težak, M. Čolić, V. Hrust, S. Popović*), S. Prgomet, and F. Strajnar

Department of Physical Chemistry, Faculty of Science and *)Rugjer Bošković Institute, University of Zagreb, Zagreb, Yugoslavia

Abstract: Precipitation processes, solubility and phase behaviour in aqueous solutions of ferric nitrate − dodecyl benzene sulphonic acid (HDBS) − nitric acid, were investigated in various pH regions. Precipitation diagrams were made using light scattering. The solubility product and the enthalpies of precipitation were calculated taking into account the hydrolytic constants of Fe^{+3}, $Fe(OH)^{+2}$, $Fe(OH)_2^+$ species and the dissociation constant of HDBS. The solubility product of ferric dodecyl benzene sulphonate ($Fe(DBS)_3$) was calculated to be $K_s^0 = (1.23 \pm 0.80) \times 10^{-24}$ for pH 2.2 at 293 K. The slope of the straight line from which the solubility product was calculated, (0.317) is close to the theoretical value for 3:1 electrolyte.

Using a polarization microscope and X-ray diffraction different phases were characterized: the solid crystal, the lamellar liquid crystal and the mixtures of solid and liquid crystalline phases. The enthalpies of crystallization and liquid crystal formation were determined as $\Delta H = -33 \text{ kJ mol}^{-1}$ and $\Delta H = 23 \text{ kJ mol}^{-1}$, respectively.

Key words: Phase behaviour, liquid crystals, solubility product

Introduction

In the electrolytic solutions of dodecyl benzene sulphonic acid (HDBS) the formation of different kinds of heterogeneous phases has been found recently: solid crystals, liquid crystals and their mixtures, depending on the concentrations of the reacting components [1, 2]. The purpose of the investigations described in this paper was to examine the precipitation conditions in $Fe(NO_3)_3 - HBDS - HNO_3$ aqueous solutions. The construction of the precipitation boundary diagrams is of interest for detergency studies [3]. In this work the formation of $Fe(DBS)_3$ solid phase was followed by light scattering. Inside the precipitation diagram (in the equivalency region, as well as in the great excess of the reacting ions) the characteristics of the precipitates were followed by polarization microscopy and X-ray diffraction. The hydrolysis of Fe(III) in aqueous solutions occurs even at pH 1 in several steps; the simultaneous presence of mono- [4, 5, 6] and polynuclear complexes, as well as polymers [7 − 10] has been found. It was possible to calculate the solubility product of $Fe(DBS)_3$ under controlled experimental conditions.

Experimental

Materials

HDBS was supplied by "Prva Iskra" (Barić, Beograd) and was a mixture of branched chain acids (benzene sulphonic group on $C_1 - C_4$ according to NMR analysis) up to 97−98%: sulphonic acid 1% and a nonsulphonized part, 1.5−2%. The molar concentration of HDBS was calculated as a nominal value of a molecular mass of 326.5. The homogeneous solution was prepared by dissolving the commercial detergent in double distilled water. The standardization was carried out potentiometrically. The titration was run by "on line" (Apple II+) potentiometric titration with standard NaOH solution; pH was monitored using an "Iskra" glass electrode. A calomel electrode served as a reference with a bridge of $NaNO_3$ solution (10^{-3} mol dm^{-3}) [11].

The stock solution of 1 M ferric nitrate (in water or in 10^{-2} mol dm^{-3} HNO_3) was prepared by dissolving $Fe(NO_3)_3 \cdot 9H_2O$ from "Kemika" (Zagreb), in double distilled water and standardized by EDTA with Variamine Blue as indicator [12]. All the diluted concentrations were freshly prepared just before experimentation.

Methods

Precipitation systems were prepared by mixing the reacting components by the method described previously [1, 13].

The methods of sample preparation for the light microscope and for X-ray diffraction have been also described [1]. The calorimetric analysis were carried out with a reaction calorimeter, described by Simeon [14]. The samples for calorimetric measurements were prepared by adding (by microburette) 1.03 ml of 1 mol dm^{-3} solution of Fe(NO$_3$)$_3$ (final concentration 1.5×10^{-2} mol dm^{-3}) to the reaction vessel, containing 70 cm^3 of HDBS solution 10^{-2} mol dm^{-3} (sample 1, Fig. 3); or by adding 1.47 ml of 0.25 mol dm^{-3} solution of HDBS (final concentration 5.25×10^{-3} mol dm^{-3}) to the reaction vessel containing 70 cm^3 of Fe(NO$_3$)$_3$ 4×10^{-1} mol dm^{-3} (sample 2, Fig. 3). The extent of reaction was calculated by assuming the presence of hydrolytic species from [15−18] and by using the solubility constant obtained in this work.

Techniques

Light scattering determination of precipitation/solubility boundary, i.e. the determination of the "first clear" system 1 h after mixing the reacting components, was performed by universal light scattering photometer Virtis Brice Phoenix DU 2000 and by Zeiss tyndallometer connected to a Pulfrich photometer.

The c.m.c. was determined by light scattering. The UV spectra of Fe(III) solutions were obtained on Pye Unicam SP 1800 spectrophotometer. pH measurements were made on a pH-meter, Philips PW 2419/18.

The characterization of solid and liquid crystalline phases was carried out using a Leitz Wetzlar light microscope with polarization equipment and by X-ray diffraction.

Micellization and dissociation of HDBS

The micellization of HDBS in aqueous solutions has been previously investigated [1], but the effects are different in electrolytic and acidic medium. Figure 1 shows the shift of c.m.c., when the ionic strength or pH are changed. The degree of dissociation of HDBS in aqueous and nitric acid solutions has been determined by "on line" potentiometric titration. The degrees of dissociation (a) are tabulated in Table 1. In the data treatment for the solubility product

calculation, it was assumed that HDBS was totally dissociated under the given conditions.

Interpretation of data and calculations

The equilibrium at the solubility limit at 20°C, with controlled pH 2.2, can be represented by the reaction:

$$Fe(DBS)_3 \rightleftharpoons Fe^{+3} + 3\,DBS^- \qquad (1)$$

The solubility product is therefore:

$$K_s^0 = a(Fe^{+3}) \cdot a^3(DBS^-), \qquad (2)$$

and can be expressed as:

$$K_s^0 = c(M) \cdot y(M) \cdot c^3(D) \cdot y^3(D), \qquad (3)$$

where activity coefficient $y(M)$ can be assumed to be:

$$y(M) \equiv y^9(D) \equiv y^9; \qquad (4)$$

and calculated using Debye-Hückel law. The ionic strength is calculated by taking into account all ionic species which are present in solution. It is known that in the iron solutions at a pH region of 1.5−2.5, equilibria exist with the hydrolytic constants [15−18]:

$$Fe^{+3} + H_2O \rightleftharpoons Fe(OH)^{+2} + H^+; \qquad k_{11} = 3.69 \cdot 10^{-3} \quad (5)$$

$$Fe^{+3} + 2H_2O \rightleftharpoons Fe(OH)_2^+ + 2H^+; \qquad k_{21} = 4.9 \cdot 10^{-7} \quad (6)$$

$$2Fe^{+3} + 2H_2O \rightleftharpoons Fe_2(OH)_2^{+4} + 2H^+; \qquad k_{22} = 1.4 \cdot 10^{-3}. \quad (7)$$

In the HDBS solution, the equilibrium takes place with the relevant dissociation constant

$$HDBS \rightleftharpoons H^+ + DBS^-; \qquad k_a = 2.02 \cdot 10^{-2}. \qquad (8)$$

Therefore the total concentration of Fe(III) ions is:

$$[Fe^{+3}]_{TOT} = [Fe^{+3}] + k_{11}\frac{[Fe^{+3}]}{[H^+]} + k_{21}\frac{[Fe^{+3}]}{[H^+]^2} + 2k_{22}\frac{[Fe^{+3}]^2}{[H^+]^2} \qquad (9)$$

The final expression for the K_s^0 is:

$$K_s^0 = c(M) \cdot c^3(D) \cdot y^{12}.$$

This equation was used for linear fitting.

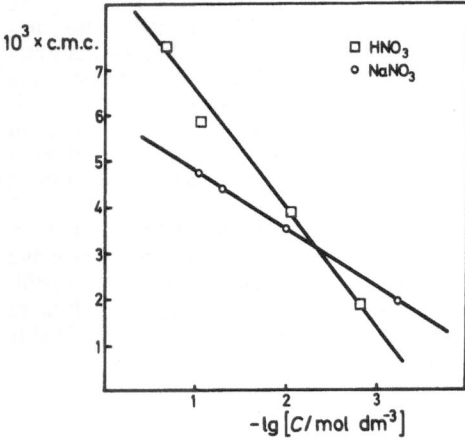

Fig. 1. Linear shift of c.m.c. of HDBS in HNO$_3$ and NaNO$_3$ solutions

Table 1. The degree of dissociation of HDBS in aqueous or acidic solutions

c(HDBS) mol dm^{-3}	$a/\%$ (in water)	$a/\%$ (in 0.01 molar HNO$_3$)
10^{-2}	95	88
10^{-3}	98.1	91.2
10^{-4}	99.1	92
10^{-5}	99.995	93.5
10^{-6}	99.9999	95

Results and discussion

The precipitation systems were prepared by the mixing of ferric nitrate and HDBS solutions (with or without HNO_3). The samples were prepared in the equimolar ratio, and with great excess of one of the reacting components. The contours of solubility boundary are presented in Fig.2. It is a typical precipitation diagram of metal-detergent in aqueous solution [2], which shows various regions of precipitation. At low concentrations of detergent with excess metal ion, the boundary is parallel with the abscissa, giving an intercept with the equimolarity line at nearly equimolar concentrations (solubility product region), and the complex ion formation in the detergent concentrations higher than c.m.c. The slope of the straight line of the solubility limit in the equimolarity region of 0.317, fits the theoretical slope for 3:1 electrolytes at pH 2.2. It is obvious that the calculation of the solubility product could have been made only in this equimolarity region for the experimental points lying on the straight line. The solubility boundary for systems prepared without HNO_3 (curve 1) results in the slope fitting approximately 1:1 electrolytes (pH 4—5, where uni- and bivalent hydrolytic species exist in solution). The presence of 0.25 M HNO_3 (curve 3) causes higher solubility than the presence of 0.01 M HNO_3 (curve 2) because of decreased dissociation degree of HDBS. The slope of the straight line intercepting the equimolarity line is, for these systems, also 3:1, due to the predominant presence of Fe^{+3}

ions. Under strictly controlled conditions (20 °C, pH = 2.2, a = 95%, taking into account that Fe^{+3} species were present at 78%) K_s^0 was calculated as $(1.23 \pm 0.80) \times 10^{-24}$.

In Fig. 2 the large open circles denote the samples that were chosen for X-ray examinations.

Investigation of the precipitated phase behaviour in the higher regions of the concentration of both reacting components, was performed using the microscope in polarized light. Different regions of precipitates are shown in Fig. 3. The formation of solid crystal (SC), smectic A lamellar mesophase (LC) and mixtures of solid and liquid crystals with a predomination of solid (SC+LC) or liquid crystal phase (LC+SC) were found. Liquid crystals show the optical birefringence, exhibiting the symmetrical cross typical for smectic A mesophases. In Fig. 4, three samples are presented: (a) the sample of SC (opaque crystals in the polarized light) with a very small appearence of birefringent LC, (b) the "oily streaks" appearence of the LC phase and (c) single droplets of LC phase.

The droplets were formed in chains, thus exhibiting the "oily streaks" appearence. The formation of LC phase was mainly found in the wide equivalence of reacting components. It was mostly expressed in the samples without HNO_3 (Fig. 3a) in the concentrations of HDBS higher than c.m.c. The presence of 0.25 M HNO_3 (in Fig. 3c) in the systems, causes change in the degree of HDBS dissociation and micellization, therefore (c.m.c.)* is only a nominal

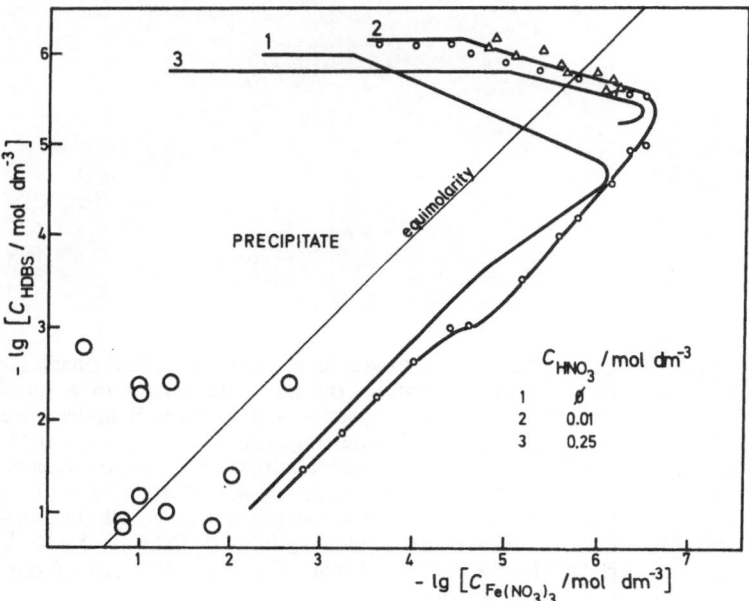

Fig. 2. Precipitation diagram $Fe(NO_3)_3$-HDBS with the contours of the solubility boundary in the systems: (1) without HNO_3, (2) with 0.01 = M HNO_3, (3) with 0.25 M HNO_3. The open circles in the high concentrations denote the samples chosen for X-ray analysis. The signs $\triangle$ or $\circ$ denote the points from two sets of experimental determination of solubility limit

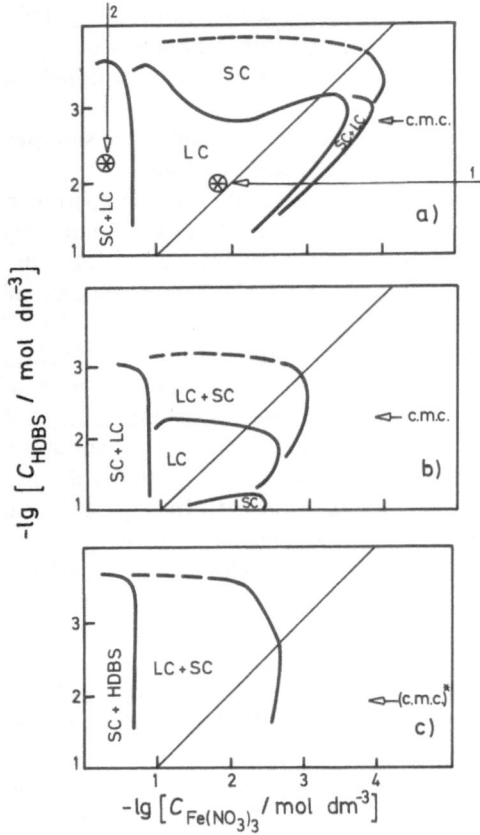

Fig. 3. The regions of formation of various phases: (a) without HNO₃; the arrows denote the samples chosen for calorimetric measurements, (b) with 0.01 M HNO₃, (c) with 0.25 M HNO₃

value, and there is no appearance of "pure" LC phase. Solid crystal formation occurs mainly in the great excess of Fe^{+3} ion, due to the higher rate of nucleation, i.e. the crystal nuclei are formed in great numbers, the crystals grow fast and there is no time for LC phase formation.

4b

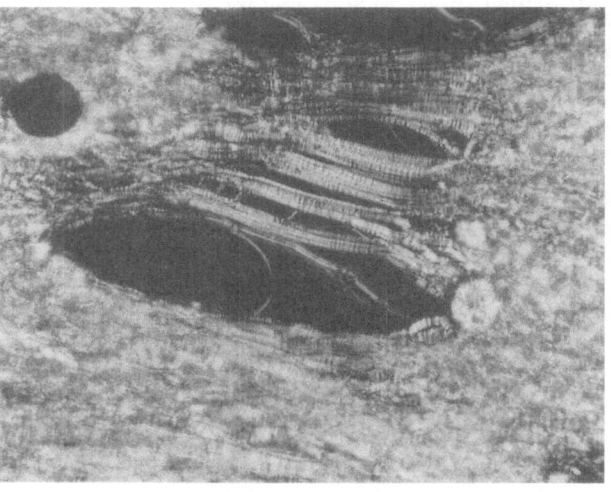

4c

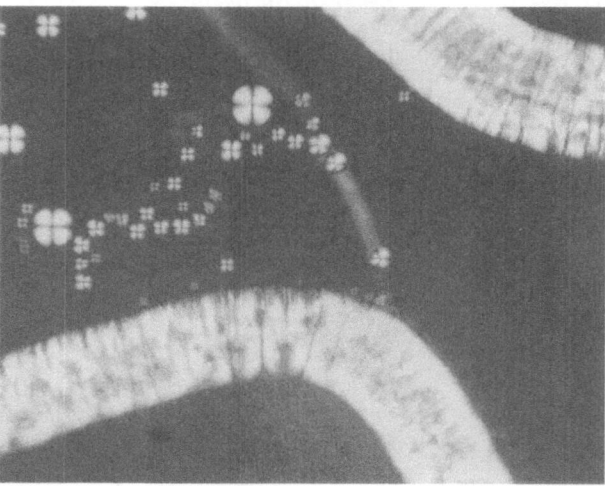

Fig. 4. The light micrographs of $Fe(NO_3)_3$-HDBS precipitation systems presenting: (a) the solid crystal in a great predominance (dark crystals) with the small appearence of liquid crystals (birefringent); [$Fe(NO_3)_3$] = 4×10^{-1}, [HDBS] = 2.5×10^{-2}; total magnification ×200; (b) viscous mass and "oily streaks" liquid crystals: [$Fe(NO_3)_3$] = 10^{-1}, [HDBS] = 2.5×10^{-2}; total magnification ×200; (c) droplets and chains of liquid crystals: [$Fe(NO_3)_3$] = 5×10^{-3}, [HDBS] = 6×10^{-3}; total magnification ×500. All concentrations in mol dm⁻³

4a

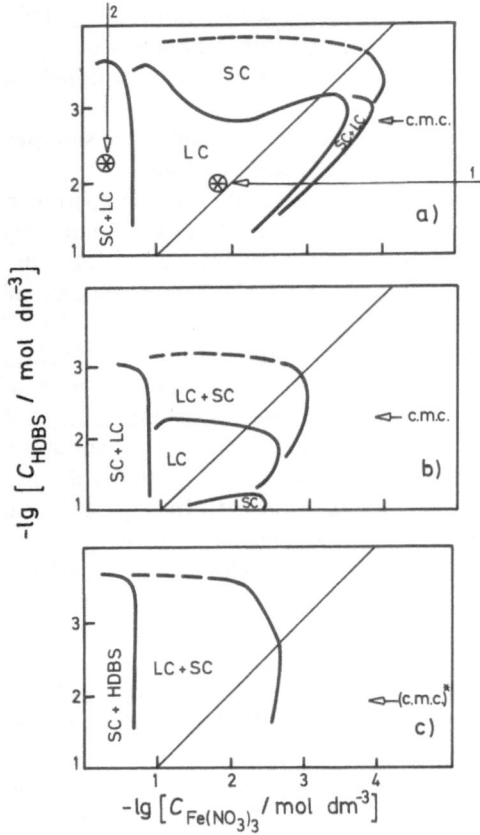

The enthalpy of solid crystal formation is found to be $\Delta H = -33\,\mathrm{kJ\,mol^{-1}}$ for the sample denoted by arrow 2. Liquid crystal formation is an endothermic process of $\Delta H = 23\,\mathrm{kJ\,mol^{-1}}$ (sample denoted by arrow 1).

The X-ray analysis of the samples denoted in Fig. 2 by large open circles shows the existence of lamellar mesophase, as found previously [1, 2] for other metal-detergent systems. Figure 5 is one of representative diffraction patterns (monochromatized CuK_α radiation) showing three orders of reflection from one basic lamellar thickness D, corresponding to the spacings 32.5, 16.2, 10.8 and 8.0 Å. The basic lamellar thickness D for the samples with 0.01 M HNO_3 was found to be on average 1.5 Å higher than in the samples without HNO_3:

$$D\ (\text{without}\ HNO_3) = (31.1 \pm 0.6)\ \text{Å},$$

$$D\ (\text{with}\ HNO_3)\quad = (32.6 \pm 0.4)\ \text{Å}.$$

It can be assumed that the increase of D in the presence of a higher concentration of H_3O^+ ions is due to the change in the depth of the hydration shell of Fe^{+3} species in the polar part of the lamellae.

Conclusion

To characterize the heterogeneous processes in ferric-detergent systems, various equilibria have to be taken into account, including hydrolysis of $Fe(NO_3)_3$-solutions, as well as micellization and dissociation in detergent solutions. The formation of solid and liquid crystalline phases is strongly dependent on the c.m.c. of HDBS, as well as on the pH of the medium. The enthalpies of crystal and liquid crystalline phase formation can be easily distinguished as exothermic and endothermic processes, respectively. The liquid crystals are formed in the equivalency region, in the concentrations higher than c.m.c. of HDBS. The relative lack of HDBS in the excess concentrations of $Fe(NO_3)_3$ creates the conditions for the very fast nucleation and growth of solid crystals.

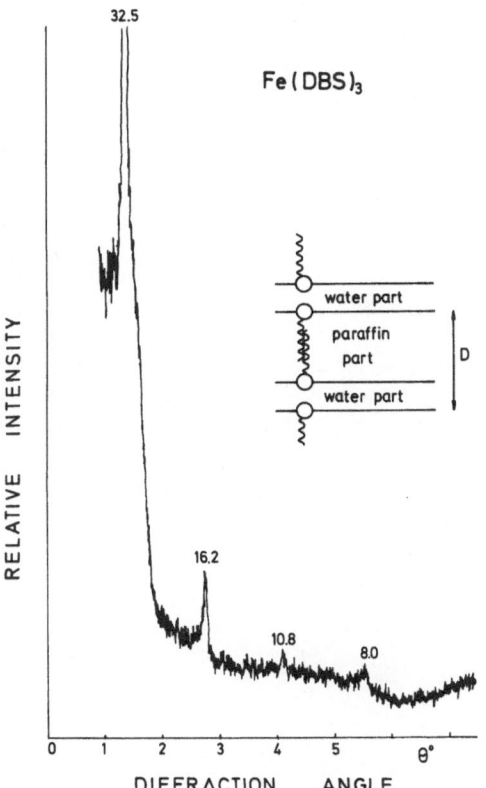

Fig. 5. X-ray diffraction at small angles gives the typical periodical reflections ($d = 32.5$, 16.2, 10.8 Å, respectively) of smectic liquid crystals, and the reflection at 8.0 Å. Concentrations in the sample are in mol dm^{-3}: [Fe(NO$_3$)$_3$] = 10^{-1}, [HDBS] = 6×10^{-2}, [HNO$_3$] = 10^{-2}. The schematic presentation shows the unit layer composed of paraffin and water in the layer thickness D.

References

1. Težak D, Strajnar F, Šarčević Đ, Milat O, Stubičar M (1984) Croat Chem Acta 57:93
2. Težak Đ, Strajnar F, Milat O, Stubičar M (1984) Progr Colloid Polym Sci 69:100
3. Matheson KL (1985) J Am Oil Chem Soc 62:1269
4. Musić S, Vértes A, Simmons GW, Czako-Nagy I, Leidheiser H, Jr (1982) J Colloid Interface Sci 85:256
5. Leidheiser H, Musić S, McIntyre JF (1984) Corros Sci 24:197
6. Milburn RM, Vosburgh WC (1955) J Am Chem Soc 77:1352
7. Hödstrem BOA (1953) Ark Kemi 6:1
8. Dousma J, de Bruyn PL (1978) J Colloid Interface Sci 64:154
9. Murphy PJ, Posner AM, Quirk JP (1976) J Colloid Interface Sci 56:312
10. Dousma J, de Bruyn PL (1976) J Colloid Interface Sci 56:527
11. Ćolić M (1985) BSc Thesis, Zagreb
12. Vogel AI (1961) A Textbook of Quantitative Inorganic Analysis, 3rd Edition, Longman, London
13. Težak B, Matijević E, Schulz K (1951) J Phys Colloid Chem 55:1558
14. Simeon VI, Ivičić N, Tkalčec M (1972) Z Phys Chem, Neue Folge 78:1
15. Sapieszko RS, Patel RC, Matijević E (1977) J Phys Chem 81:1061

16. Matijević E, Janauer GE (1966) J Colloid Interface Sci 21:197
17. Strahm U, Patel RC, Matijević E (1979) J Phys Chem 83:1689
18. Matijević E, Scheiner P (1978) J Colloid Interface Sci 63:509

Received October 8, 1986;
accepted October 18,1986

Authors' address:

Đ. Težak
Department of Physical Chemistry
Faculty of Science
University of Zagreb
Marulićev trg 19/II
P.O. Box 163
41001 Zagreb, Yugoslavia

Association of sodium ions to aqueous alkylsulfate and alkanoate micelles in the presence of 1-alcohols

S. Backlund[1], K. Rundt[1], K. Veggeland[2] and H. Høiland[2]

[1] Department of Physical Chemistry, Åbo Akademi, Åbo, Finland
[2] Department of Chemistry, Bergen University, Norway

Abstract: Electromotive force and conductivity measurements have been used to study the counter ion association to micelles of sodium dodecyl sulfate, sodium dodecanoate, and sodium octanoate. The conductivity and electromotive force measurements match very well as both is mainly a measure of the amount of free counter ions in the solutions. Addition of alcohols to these micellar solutions show some differences between octanoate micelles on one hand, and dodecyl sulfate and dodecanoate on the other. Propanol and pentanol addition lead to an increase in the sodium activity for dodecyl sulfate and dodecanoate solutions that are well above the c.m.c. while a decrease is observed for octanoate. Addition of higher alcohols show that the counter ion activity and the conductivity go through a maximum at low alcohol contents and then decreases. At high surfactant contents alcohol addition may induce a shape transition from small spherical micelles to larger rod- or disc-like micelles. It appears that the degree of counter ion association increases when these larger micelles form.

Key words: Micelles, counter ion association, shape transitions, 1-alcohols, sodium ion activity, conductivity, viscosity.

Introduction

When ionic surfactants aggregate to form micelles, the counter ions will be partly associated with the micelles and partly dispersed in the aqueous surroundings. The fraction of counter ions associated with the micelles, β, has previously been determined by several experimental techniques such as light scattering, NMR, conductivity, and electromotive force measurements [1, 2]. In most cases β is found in the range 0.5 to 0.8 [3].

An important property of micellar solutions is the ability to solubilize a third component that is otherwise sparingly soluble in water. Solubilization will affect the micellar charge density and hence the counter ion association. It is normally reduced when alcohols become solubilized, while there is little change for the solubilization of alkanes. Furthermore, solubilization can lead to structural changes in micelles from small spherical to larger rod- or disc-like micelles. This may also affect the counter ion association. A small increase in the counter ion association has been observed for hexadecyltrimethylammonium bromide [4].

Carboxylate micelles appear to behave differently with respect to the counter ion association when compared to sodium dodecyl sulfate. Solubilization of

decanol in sodium octanoate increases the counter ion association [4]. The EMF measurements of Vikingstad [5] show that addition of octanol and decanol also increase the counter ion association in sodium decanoate. Solubilization data for hexanol in sodium dodecyl sulfate, hexadecyltrimethylammonium bromide, and sodium decanoate, indicate that the entropy of solubilization is almost twice as large for the decanoate system [6].

The results so far indicate a difference between carboxylate and sulfate micelles that affects the counter ion association when a medium chain length alcohol becomes solubilized. In order to elucidate this possible difference we have measured the sodium ion activity and the electrical conductivity of sodium octanoate, dodecanoate and dodecyl sulfate as 1-alcohols are added.

Experimental

Sodium dodecyl sulfate (NaDDS) was supplied by BDH, of „specially pure" grade. Sodium octanoate (NaC$_8$) and sodium dodecanoate (NaC$_{12}$) were prepared from the corresponding acids (Fluka) as previously described [7]. The alcohols, 1-propanol, 1-butanol, 1-pentanol, 1-octanol, and 1-decanol, were supplied by Fluka at their best quality.

Sodium chloride was used as supplied by E. Merck (p.a. quality). Water was distilled and passed through an ion exchange resin immediately before use. The conductivity of water was 0.5 µS cm^{-1}. All solutions were made by weight.

The electromotive force measurements were made with a sodium responsive glass electrode type GEA 33, manufactured by Electronics Instruments Ltd. The reference electrode was a double junction electrode, model 90−92 from Orion with a 3 M NH$_4$NO$_3$ bridge solution. The details of the cell and measuring procedure have been described elsewhere [8]. The electrical conductivities were measured by a Wayne-Kerr bridge with automatic recording of the resistances [9]. The solution densities were measured by a Paar density meter and the viscosities by Ostwald viscometers. All solutions were Newtonian. The temperature was kept at 298.15±0.01 K.

Results and discussion

Nernst's equation for the cell is given by:

$$E = E_j^0 - k_E \log a_{Na^+} . \tag{1}$$

E is the measured potential and E_j^0 a term containing both the standard potential and the liquid junction potential. The Nernstian slope, k_E, was close to the theoretical value for NaCl solutions. By determining $\Delta E = (E^0 - E)$, where E^0 and E are the measured cell potential in alcohol free and alcohol containing solutions, the relative sodium ion activity $a_{Na^+}/a_{Na^+}^0$ can be determined from:

$$\Delta E = k_E \log (a_{Na^+}/a_{Na^+}^0) . \tag{2}$$

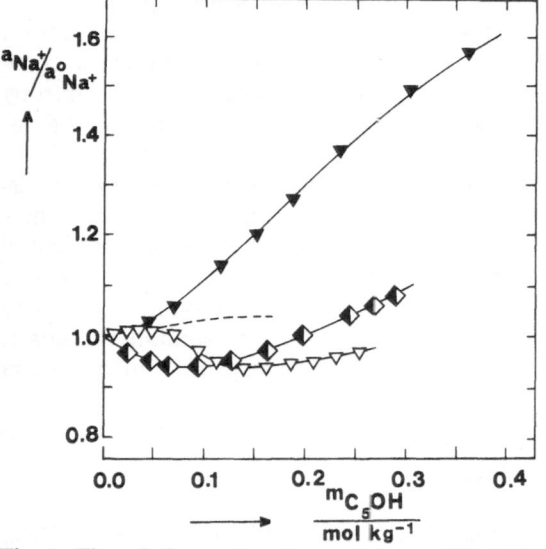

Fig. 1. The relative sodium ion activity at 298.15 K in sodium dodecyl sulfate solutions as a function of added pentanol. The sodium dodecyl sulfate molalities are: ($\triangledown$) 3.1×10^{-3}, ($\blacklozenge$) 9.9×10^{-3}, ($\blacktriangledown$) 3.7×10^{-2} m. The dashed line is 0.1 m NaCl

Figure 1 shows the effect to pentanol on the sodium ion activity for three different NaDDS solutions. It also shows how the addition of pentanol affects the measured potential in a solution of 0.1 m NaCl. When pentanol is added to a 0.0031 m NaDDS solution (well below the c.m.c. in water), the sodium activity initially varies, like that of NaCl. As more pentanol is added, the sodium ion activity decreases, indicating that micelles are forming and that part of the sodium ions become associated with the micelles. At even higher pentanol contents the sodium ion activity starts to increase again. At this point, micelle formation is no longer the important factor. Now one observes dissociation of counter ions due to pentanol solubilization. The overall variation of the sodium activity of this NaDDS solution upon pentanol addition is small, only about 4%. At higher NaDDS contents, smaller amounts of pentanol are needed to induce micelle formation, as expected. For solutions that are already above the c.m.c., only an increase in the sodium activity is observed since solubilization of pentanol is the major effect. This will lower the micellar charge density and release associated counter ions. Sodium dodecanoate solutions show the same general behavior as NaDDS solutions with regard to the sodium ion activity as pentanol is added. However, sodium octanoate is different: we observe a decrease in the sodium ion activity in all cases.

Changes in the counter ion association upon solubilization of alcohols also affect the electrical conductivity of the solutions [10, 11]. Figure 2 shows the relative conductivity for the same systems as in Fig. 1. The curves match very well.

Vikingstad [5] and Tominaga et al. [11] have shown that the addition of alcohols affects the content of free counter ions far more than the content of free surfactant monomers. This means that the conductivity is basically a measure of the amount of unassociated counter ions. This can be seen from the following, since the conductivity of a surfactant solution containing i different ionic species is:

$$\kappa \sim \sum_i z_i^2 c_i/r_i . \tag{3}$$

Here z_i, r_i and c_i are the charge, radius, and concentration of one ionic species. Below the c.m.c., where $z_i = 1$ and c_i is the total concentration, one will obtain the following proportionality:

$$\kappa \sim c_{tot}(1/r_+ + 1/r_-) . \tag{4}$$

Usually r_+, i.e. the radius of the counter ion is much smaller than r_- the radius of the surfactant mono-

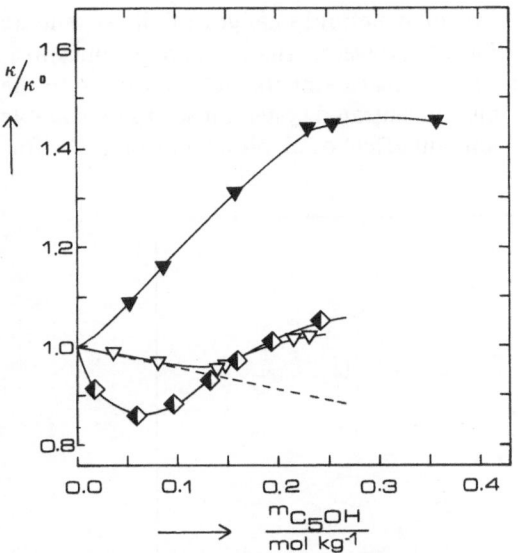

Fig. 2. The relative conductivity at 298.15 K of sodium dodecyl sulfate solutions as a function of added 1-pentanol. The symbols are as in Fig. 1. κ^0 is the conductivity of the pentanol free surfactant solution

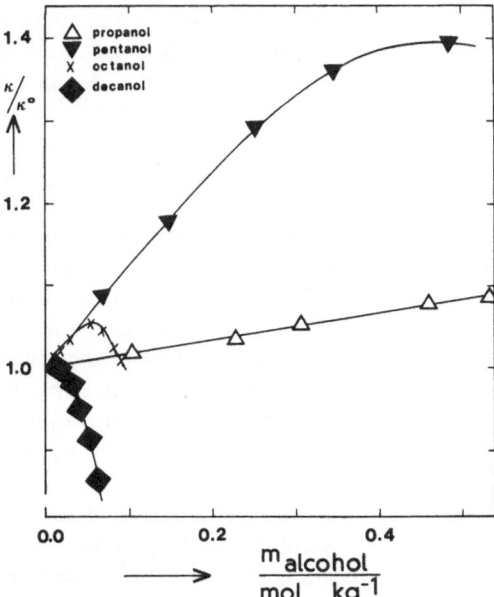

Fig. 4. The relative conductivity of a 0.176 m solution of sodium dodecyl sulfate at 298.15 K as a function of added 1-alcohols

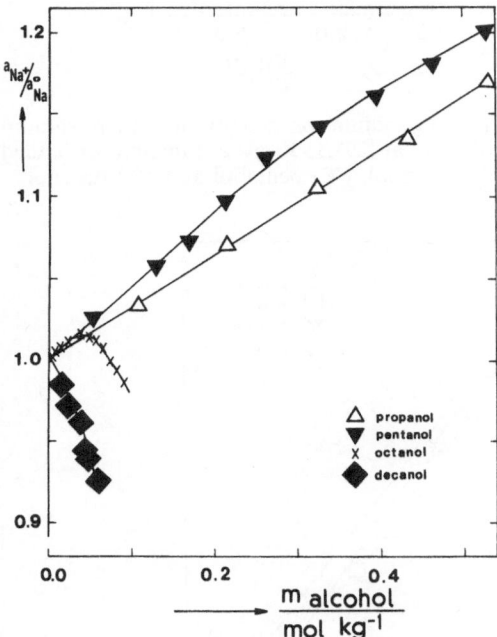

Fig. 3. The relative sodium ion activity of a 0.176 m solution of sodium dodecyl sulfate at 298.15 K as a function of added 1-alcohols

mer, and the conductivity is almost a measure of the counter ion conductivity alone. For ionic micelles the charge is still unity for the free counter ions, and it is aN for the micelle (a is the degree of counter ion

dissociation and N the micellar aggregation number). The conductivity is thus proportional to:

$$\kappa \sim \frac{c_{\text{tot}} - \beta(c_{\text{tot}} - c_c)}{r_+} + \frac{c_c}{r_-} + \frac{a^2 N(c_{\text{tot}} - c_c)}{r_{\text{mic}}} . \qquad (5)$$

c_c is the concentration at the c.m.c.. Close to the c.m.c. the difference $c_{\text{tot}} - c_c$ is almost zero and only the counter ion term contributes significantly to the conductivity. The term c_c/r_- is small and almost constant. For concentrations well above the c.m.c. the last term is not negligible, but a comparison of Figs. 1 and 2 shows that the conductivity and the counter ion activity still correlate very well.

Figures 3 and 4 show the relative counter ion activity and the relative conductivity of a 0.176 m NaDDS solution as a function of added alcohol. The activity and conductivity data correlate strongly. The counter ion activity and the conductivity increase slightly as propanol is added. This is probably just a solvent effect, since propanol is not solubilized significantly in micellar solutions [12]. Addition of pentanol also increases the counter ion activity and the conductivity, more so than propanol. In this case, the increase can be explained by a dissociation of micellar counter ions as pentanol becomes solubilized. The activity curves and conductivity curves match very well. When octanol is added, one first observes an increase in the counter ion activity matched by an increase in conductivity. This shows that counter ions are dissociated as octanol becomes solubilized by the micelles. As the octanol content increases a maximum is reached, after which both the counter ion activity and the conductivity decrease. With addition of decanol it appears that the counter ion activity and the conductivity decrease at all contents. One possible explanation for the observed decrease is structural changes of the micelles from near spherical micelles to large rod- or disc-like micelles. In order to check this, the viscosities of NaDDS solutions have been measured as a function of added alcohol. Figure 5 shows the data. It is apparent that the viscosity changes, upon addition of propanol and pentanol, are too small to justify any rod- or disc-like micelles. The linear increase observed can be explained by a small increase in the micellar volume as these alcohols become solubilized [13], or by an increasing number of micelles, since alcohols lower the c.m.c. Addition of octanol and decanol, on the other hand, leads to an exponential increase in viscosity. This can be interpreted as a transition from small, spherical micelles to larger rod- or disc-like micelles [11, 13–17]. According to Kamenka et al. [4] the degree of counter ion association increases when rod- or disc-like micelles form, and this agrees well with the present data.

Solubilization of alcohols in 1.0 m sodium octanoate solutions appear to be quite different from those of the NaDDS or NaC$_{12}$ systems. Figures 6 and 7 show that the relative sodium ion activity and the relative conductivity decrease as alcohols are added. Kamenka et al. [4] and Vikingstad [5] have observed the same effect. Alcohols appear to increase the counter ion association of octanoate micelles. Viscosi-

ty data show no structural changes of the octanoate micelles (Fig. 8). However, the surfactant concentrations of all these measurements are fairly close to the c.m.c.. In this concentrated system, it seems reasonable that the dominant effect of alcohol addition is the for-

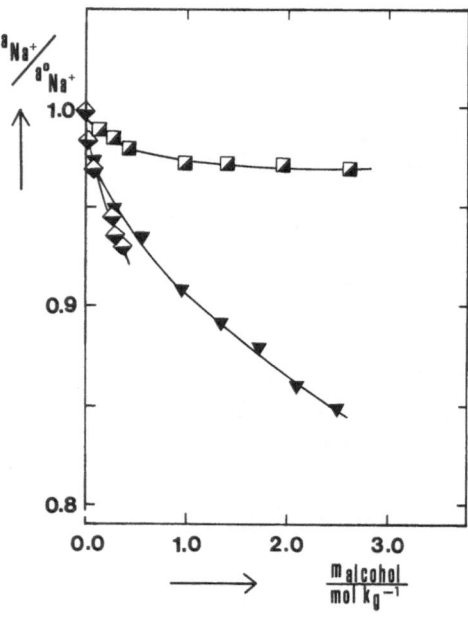

Fig. 6. The relative sodium ion activity of a 1.0 m sodium octanoate solution at 298.15 K as a function of added 1-alcohols: (□) butanol, (▼) pentanol and (◆) decanol

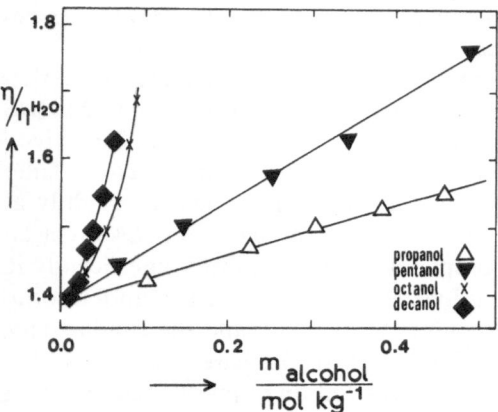

Fig. 5. The relative viscosity of a 0.176 m solution of sodium dodecyl sulfate at 298.15 K as a function of added 1-alcohols. η^{H_2O} is the viscosity of water

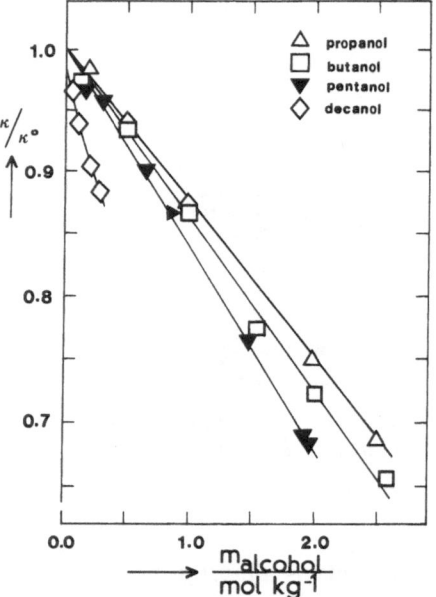

Fig. 7. The relative conductivity of a 1.0 m solution of sodium octanoate at 298.2 K as a function of added 1-alcohols

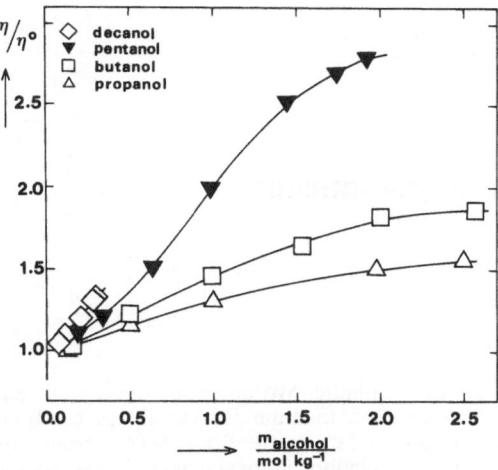

Fig. 8. The relative viscosity of a 1.0 m solution of sodium octanoate at 298.15 K as a function of added 1-alcohols. η^0 is the viscosity of the alcohol free surfactant solution

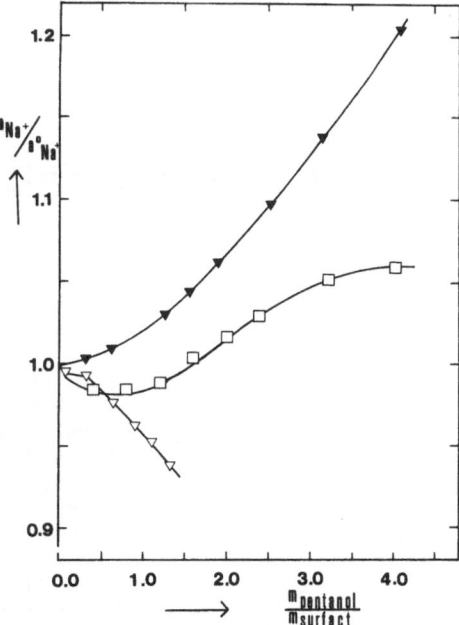

Fig. 9. The relative sodium activity of sodium dodecyl sulfate, sodium dodecanoate, and sodium octanoate solutions at 298.15 K as a function of added 1-pentanol. The surfactant molalities are all 4.6 times the molality at the c.m.c., i.e. (▼) 0.037 m NaDDS, (□) 0.124 m NaC_{12} and (▽) 1.832 m NaC_8

mation of more micelles. The more micelles, the more counter ions will be associated to micelles, and the counter ion activity and the conductivity will decrease.

Figure 9 compares the effect of pentanol on sodium dodecyl sulfate, sodium dodecanoate, and sodium octanoate where the surfactant content is equally far from the c.m.c.; i.e. 4.6 × c.m.c. Qualitatively, the counter ion activities of the dodecyl sulfate and dodecanoate solutions are similar, but NaC_8 is different. The relative sodium activity decreases as before, but much less than at the lower NaC_8 concentration in Fig. 7. It is still possible that the formation of new micelles plays a part, but this moderate decrease in sodium ion activity could well be due to the formation of rod-like micelles. Passinen and Ekwall [13] have suggested that such micelles exist in this concentration region.

Acknowledgement

The authors wish to thank Statoil (VISTA program) for financial support.

References

1. Wennerström H, Lindman B (1979) Phys Rep 52:1
2. Lindman B, Wennerström H (1980) Topics Curr Chem 87:1
3. Lindman B, Wennerström H (1982) In: Mittal KL, Fendler EJ (eds) Solution Behavior of Surfactants, vol 1. Plenum Press, New York, p 3
4. Kamenka N, Fabre H, Chorro M, Lindman B (1977) J Chim Phys 74:510
5. Vikingstad E (1980) J Colloid Interface Sci 73:260
6. Høiland H, Blokhus AM, Kvammen OJ, Backlund S (1985) J Colloid Interface Sci 107:576
7. Høiland H, Kvammen OJ, Backlund S, Rundt K (1984) In: Mittal KL, Lindman B (eds) Surfactants in Solution, vol 2. Plenum Press, New York, p 949
8. Backlund S, Rundt K (1980) Acta Chem Scand A34:433
9. Blokhus AM, Høiland H, Backlund S (1986) J Colloid Interface Sci 114:9
10. Lawrence ASC, Pearson JT (1967) Trans Faraday Soc 63:495
11. Tominaga T, Stem TB, Evans DF (1980) Bull Chem Soc Jpn 53:795
12. Birdi KS, Backlund S, Sørensen K, Krag T, Dalsager S (1978) J Colloid Interface Sci 66:118
13. Passinen K, Ekwall P (1956) Acta Chem Scand 10:215
14. Larsen JW, Magid LJ, Payton V (1973) Tetrahedron Lett 29:2663
15. Clarke DE, Hall DG (1974) Colloid Polym Sci 252:153
16. Hirsch E, Candau S, Zana R (1984) J Colloid Interface Sci 97:318
17. Ljosland E, Blokhus AM, Veggeland K, Backlund S, Høiland H (1985) Progr Colloid Polym Sci 70:34

Received October 14, 1986;
accepted November 6, 1986

Authors' address:

S. Backlund
Department of Physical Chemistry
Åbo Akademi
SF-20500 Åbo (Finland)

Progress in Colloid & Polymer Science   Progr Colloid & Polymer Sci 74:98–102 (1987)

Enzymatic transesterification of a triglyceride in microemulsions

K. Holmberg and E. Österberg

Berol Kemi AB, Stenungsund, Sweden

Abstract: Microemulsions based on aliphatic hydrocarbon, surfactant and aqueous buffer have been used as reaction medium for the lipase catalyzed transesterification of a triglyceride and a fatty acid. Both AOT (sodium bis (2-ethylhexyl)sulfosuccinate) and certain alcohol ethoxylates could be used as surfactant to produce a triglyceride having a fatty acid composition similar to that of natural cocoa butter from a palm oil distillation fraction. The non-ionic surfactant gave a higher reaction rate than AOT, presumably due to a more favourable association of water in the microemulsion. Recovery of the enzyme was facile with the former surfactant. However, the ethoxylate was found to participate in an unwanted side reaction, viz, formation of esters with free fatty acids in the solution.

Key words: Microemulsion, lipase, triglyceride, palm oil, cocoa butter

Introduction

Several of the normal hydrolytic enzymes have been found to catalyze also the reverse reaction, i.e. condensation. Lipase catalyzed esterification of an alcohol and an acid can, for instance, be made to proceed almost quantitatively in an essentially non-aqueous medium [1].

The enzymes often have a more rigid conformation in organic media than in water. This is reflected in an increase in thermal stability and substrate specificity. Since these properties are highly interesting from both a scientific and a commercial point of view, the field of biocatalysis in organic media is under rapid expansion at present.

Relatively recently it has been demonstrated that microbial lipases in a practically water-free medium can be used as catalysts for the transesterification of triglycerides [2–5]. Aliphatic hydrocarbons, either pure solvents, such as hexane, or distillation fractions, e.g. petroleum ether 60–80, have been found to be useful as reaction media. A small amount of water is needed, however, and this can be incorporated by the use of solid particles, e.g. a diatomaceous earth powder, on which the enzyme is precipitated and which is subsequently hydrated. The water of hydration will enable the enzymes to attain the conformation needed. Special attention has been directed

towards the production of a triglyceride mixture corresponding to cocoa butter from inexpensive triglycerides, such as palm or olive oil distillation fractions.

A typical triacylglycerol content of palm oil midfraction, as well as of cocoa butter, is given in Table 1. As can be seen, a conversion of the former into the latter would require a partial replacement of palmitoyl (P) groups by stearoyl (S) groups in 1(3) position while leaving the 2-position essentially unaffected. This has been accomplished by transesterification of a mixture of stearic acid and a distillation fraction of palm oil using a 1(3)-specific lipase as a catalyst [6–8].

Table 1. Triacylglycerol composition of a palm oil midfraction and of cocoa butter. S,P,O,L and Sat represent stearoyl, palmitoyl, oleoyl, linoleoyl and saturated acyl, respectively. (From Ref [9])

Triacylglycerol	Palm oil (%)	Cocoa butter (%)
SatSatSat	2	1
POP	65	16
POS	15	41
SOS	2	27
SatLSat	5	8
Others	11	7

In the lipase-catalyzed transesterification process, the reaction takes place at the interface between the hydrocarbon and the aqueous phases. The size of the interfacial area has been found to be rate limiting and various approaches have been taken to increase the region of contact [9]. Here, we report the use of a microemulsion as a reaction medium for the enzymatic process.

Experimental

All enzymatic reactions were carried out at 35 °C in 100 ml bottles using a magnetic stirrer. Unless otherwise stated, the reaction mixture consisted of (in parts by weight): hydrocarbon 90, palm oil fraction 5, stearic acid 5, surfactant 2 and aqueous buffer 1. 34 mg lipase was used in each run.

The lipase used was from Rhizopus sp., 50 units/mg, purchased from Sigma. Sodium bis (2-ethylhexyl)sulfosuccinate, AOT, was from Merck, Darmstadt, F.R.G. The commercial grade non-ionic surfactants were all alcohol ethoxylates from Berol Kemi AB, Stenungsund, Sweden. The pure homologues used in the model experiments were synthesized via successive additions of di- or triethylene glycol units, protected in one end, to the alcohol moiety, activated by methylsulfonyl groups. After each addition, the reaction product was deprotected and activated for the next reaction step. The final products were purified by chromatography [10].

The palm oil and stearic acid were the products of Aarhus Oliefabrik, Aarhus, Denmark.

After the reaction was complete the enzyme was denatured by heating to 90 °C for 10−20 min. The solvent mixture was evaporated under vacuum, the residue dissolved in a small amount of diethylether and the solution was loaded on preparative thin layer plates. The plates were eluated with a solvent system of petroleum ether 60−70, diethylether, formic acid (70:30:0.2 by vol). Triglyceride on the plate was visualized under UV light by spraying 0.1% 2′,7′-dichlorofluorescein and extracted with diethylether from the silica gel scraped off from the plate. The ether solution was evaporated to dryness and the triglyceride thus extracted (50−100 mg) was treated with 15 ml 0.5 M Li-methoxide for 5 min at 100 °C under gentle stirring. The solution was cooled to room temperature and 5 ml 6 M H_2SO_4 was added. After addition of 50 ml deionized water, the solution was extracted with two 35 ml portions of dichloromethane. The organic phases were collected, dried and evaporated to 10 ml volume. The solutions were analyzed for fatty acid methylesters by gas chromatography, using commercial fatty acid methylesters as standards. The proportion of stearic acid incorporated in the triglyceride was calculated from the integrals of the chromatograms.

The ester between fatty acid and surfactant was isolated and analyzed by the same procedure as used for the triglyceride fraction. In the eluent system used, the components had the following R_f-values: fatty acid ester of surfactant 0.6−0.7, triglyceride 0.5−0.6, free fatty acid 0.3 and diglyceride 0.1.

Results and discussion

1. Systems based on AOT

A considerable amount of work has been done, by Luisi and others, to explore the use of reverse micelles based on AOT (sodium bis(2-ethylhexyl)sulfosuccinate) in hydrocarbon in enzymatic processes [11−13]. Many enzymes have been found to work well in these systems, sometimes showing turnover numbers almost as high as those obtained in aqueous solution.

As can be seen in Fig. 1, the AOT-hydrocarbon-water system can be used for the lipase-catalyzed transesterification of palm oil with stearic acid. The reaction is monitored by measuring the total amount

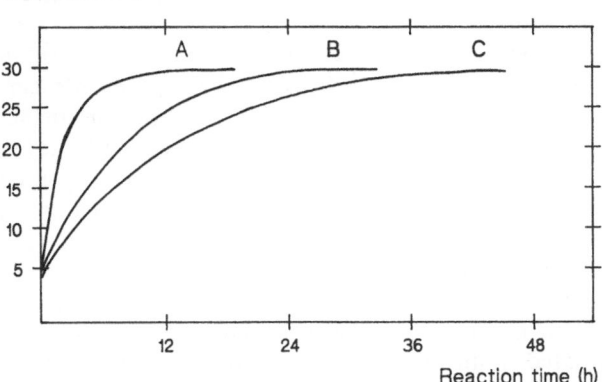

Fig. 1. Reaction rate for lipase-catalyzed transesterification of palm oil midfraction with stearic acid. Reaction rate is expressed as percentage stearic acid incorporated in the triglyceride fraction vs reaction time. (A) Microemulsion based on non-ionic surfactant. (B) Microemulsion based on AOT. (C) Hydrated celite dispersed in hydrocarbon

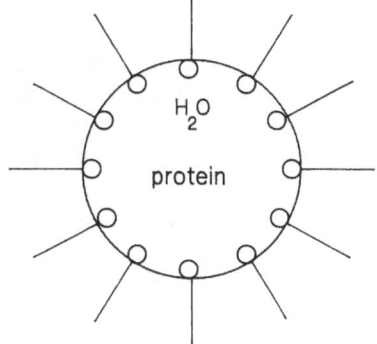

Fig. 2. The water shell model for the structure of enzymes in reverse micelles

of stearic acid in the triglyceride. The starting value, i.e. the stearoyl content of palm oil, is 6–7%. The value aimed for is of the order of 30%. The reaction is faster in the AOT system than in the heterogeneous mixture, but the increase in reaction rate is not as large as expected. The reason why the very large increase in interfacial area is not accompanied by a substantial increase in rate of reaction, could be that the enzyme is forced away from the interfacial region by the surfactant. Lipases are known to exert their action at interfaces. If they are largely located in the interior of the water pools, in accordance with the "water shell model" [14] (see Fig. 2), their ability to catalyze a transesterification reaction would be poor.

2. Systems based on non-ionic surfactants

Unlike ionic surfactants, non-ionic do not give reverse micelles at very low water content [15]. A small amount of water added to a non-ionic surfactant-hydrocarbon system will be associated to the polyethyleneglycol chain (apart from the small amount of water which dissolves in the pure hydrocarbon). Reverse micelles are not formed until the surfactant molecules are, in a way, saturated with water. Below a certain water level, the situation may then be visualized as enzyme molecules being dispersed in the hydrocarbon solvent by hydrated surfactants having their polar end oriented towards the protein surface. This arrangement would give the enzyme better access to the hydrocarbon phase, where the reactants are situated, than in the AOT system.

A number of technical grade non-ionic surfactants were tested. In general, good results were obtained with ethoxylated medium chain length fatty alcohols. Since recovery of the enzyme turned out to be particularly straight-forward with triethylene glycol monododecyl ether (see below), this surfactant was used in further work.

3. Influence of the organic solvent

Four different hydrocarbons, n-hexane, n-heptane, n-octane and n-nonane, were evaluated as organic phase. The reactions were performed at the same hydrocarbon-water-surfactant ratios which in all cases yielded one-phase systems.

As can be seen from Table 2, there is a considerable difference in reaction rate, nonane giving the fastest and hexane the most sluggish rate of incorporation of stearoyl groups into the triglyceride. The influence of the organic solvent on the rate of reaction is not fully understood. A possible explanation is that the effect is due to differences in the type of water association;

Table 2. Amount of stearoyl in the triglyceride fraction after 4 h reaction in microemulsions based on various hydrocarbons, buffer pH 8 and triethylene glycol monododecyl ether.

Solvent	Stearic acid (%)
n-Hexane	14
n-Heptane	15
n-Octane	18
n-Nonane	28

Table 3. Amount of stearoyl in the triglyceride fraction after 8 h reaction in microemulsions based on nonane, buffer of various pH and triethylene glycol monododecyl ether.

Buffer pH	Stearic acid (%)
6	20
7	31
8	23
9	17

the less hydrophobic solvents presumably giving a higher proportion of reverse micelles at the same water concentration, thus yielding a more AOT-like system (see Fig. 2). This hypothesis will be tested experimentally.

4. Influence of buffer pH

In order to investigate the influence of pH on the rate of reaction, transesterifications were carried out using buffers of pH 6, 7, 8 and 9 as aqueous phase. As can be seen from Table 3, the best results are obtained with pH 7 buffer. This, however, is no indication of the optimal pH of the reaction since it is known that the acidity within the water domains of reverse micelles may differ considerably from the original pH value of the buffer used [16]. This discrepancy is believed to be due to an uneven distribution of protonized and deprotonized species between the phases.

Luisi et al. have proposed a method of measuring the pH in the water domains of AOT-based systems [16]. An acidity scale for the water domains was defined by measuring the ^{31}P chemical shifts of phosphate buffers. The chemical shifts in bulk water were compared to those found in reverse micelles under the assumption that the pK of the phosphate ion was the same in both systems. Inconsistent results were obtained, however, when the method was applied to the present non-ionic-based system.

5. Influence of composition

The oil rich corner of the phase diagram for the system used is shown in Fig. 3. A number of compositions on both sides of the line separating one- and two-phase regions were tested. As can be seen from Table 4, the rate of reaction seems to be relatively independent of the composition, as long as it is within the one-phase region. If the composition lies in the two-phase region, the reaction rate is drastically reduced, as expected. All reactions were carried out under stirring.

In practice, the amount of water used in the reaction should be kept low in order to avoid extensive hydrolysis of the triglyceride. In the normal range of water added, 0.5−2.0%, the formation of di- and mono-glycerides is negligible, however.

Table 4. Amount of stearoyl in the triglyceride fraction after 8 h reaction in microemulsions A, B, C and D of Fig. 3.

Composition	Stearic acid (%)
A	28
B	7
C	29
D	26

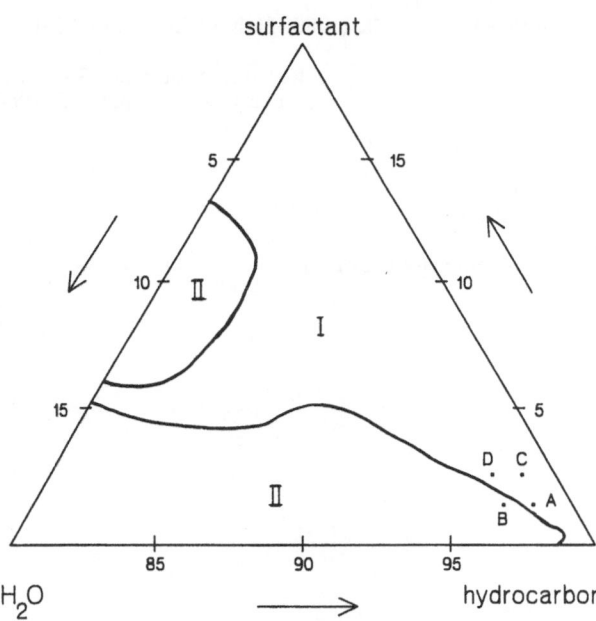

Fig. 3. Phase diagram (oil-rich corner only) for the system buffer pH 7 (H_2O), triethylene glycol monododecyl ether (surfactant) and nonane: palm oil: stearic acid (90:5:5) (hydrocarbon)

The system based on non-ionic surfactant shown in Fig. 1 has the composition of point A of Fig. 3 (1.5% surfactant, 1.0% H_2O, 97.5% hydrocarbon). It is evident that an optimized non-ionic-based system is very efficient as a medium for enzymatic transesterification of triglycerides.

6. Side reaction

The use of non-ionic surfactants with terminal hydroxyl groups gives rise to a severe side reaction, however. Under the conditions used, the enzyme has been found to catalyze not only transesterification of the triglyceride but also formation of an ester between the surfactant and free fatty acids

$$R_1O(CH_2CH_2O)_n-H+HOOCR_2 \xrightarrow{\text{lipase}}$$
$$R_1O(CH_2CH_2O)_nCOR_2$$

The ester (or, rather esters, since both the fatty acid and the surfactant consist of mixtures of species) was formed regardless of the choice of hydrocarbon and buffer pH. The amount of ester produced is normally 5−10% of the amount of triglyceride.

A commercial grade of triethylene glycol monododecylether contains a considerable amount of non-ethoxylated alcohol (see Fig. 4). Since the unreacted alcohol is more nucleophilic than the ethoxylates, it seemed possible that the byproduct mainly consisted of fatty acid esters of this compound. If this were the case, the side reaction could be minimized either by us-

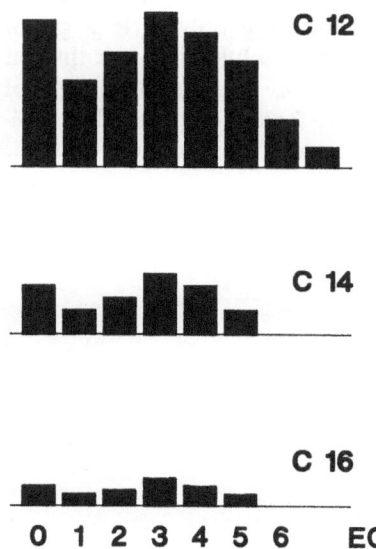

Fig. 4. Homologue distribution of the commercial surfactant triethylene glycol monododecyl ether

Table 5. Partitioning of various ethoxylated alcohols between water and nonane (1:1 by vol). R stands for dodecyl.

Ethoxylate	Water : Nonane
$R(OCH_2CH_2)_3OH$ (A)	<0.5 : >99.5
$R(OCH_2CH_2)_6OH$ (B)	8.0 : 92
A from a 1:1 mixture of A and B	0.2 : 99.8
B from a 1:1 mixture of A and B	0.3 : 99.7
A from commercial surfactant	<0.5 : >99.5
B from commercial surfactant	<0.5 : >99.5

ing an alcohol ethoxylate with a more narrow distribution of homologues or by employing an ethoxylated phenol instead. However, model experiments using pure components showed that the less reactive ethoxylated species also formed esters with fatty acids.

7. Work-up procedure

In a commercial process, recovery of the enzyme will be an important step. This is of course very easily done – simply by filtration – when using enzymes on solid particles. With triethylene glycol monododecyl ether as surfactant the recovery was found to be a simple procedure also. After the reaction is complete water is added to the reaction medium and, after phase separation, only the enzyme is found in the aqueous phase. GC analysis shows that the surfactant, despite its broad homologue distribution and its inhomogeneity of the long chain alkyl group (see Fig. 4), is entirely distributed in the organic phase.

This may seem surprising considering the fact that hydrophilic species, having long polyethylene glycol chains, are considerably more soluble in water than in hydrocarbon. For instance, hexaethylene glycol monododecyl ether in pure form is distributed to 92% into the aqueous phase, as can be seen from Table 5.

This behaviour indicates that cooperative effects are present and that surfactant aggregates are formed in the hydrocarbon phase. It must be energetically more favourable, even for hydrophilic species, to reside in such an aggregate than to dissolve in the aqueous phase.

References

1. Zaks A, Klibanov AM (1984) Science 224:1249
2. Werdelmann BW, Schmid RD (1982) Fette, Seifen, Anstrichmittel 84:436
3. Nielsen T (1985) Fette, Seifen, Anstrichmittel 87:15
4. Posorske LH (1984) J Amer Oil Chem Soc 61:1758
5. Stevenson RW, Luddy FE, Rotbart HL (1979) J Amer Oil Chem Soc 56:676
6. Macrae AR (1983) J Amer Oil Chem Soc 60:243A
7. Yokozeki K, Yamanaka S, Takinami K, Hirose Y, Tanaka A, Sonomoto K, Fukui S (1982) Europ J Appl Microbiol Biotechnol 14:1
8. Tanaka T, Ono E, Ishihara M, Yamanaka S, Takinami K (1981) Agric Biol Chem 45:2387
9. Macrae AR (1985) in: Tramper J, van der Plas HC, Linko P (eds) Biocatalysts in Organic Syntheses, Elsevier, Amsterdam, p 195
10. Andreasson E, Holmberg K (to be published)
11. Meier P, Luisi PL (1980) J Solid-Phase Biochem 5:269
12. Grandi C, Smith RE, Luisi PL (1981) J Biol Chem 256:837
13. Barbaric S, Luisi PL (1981) J Amer Chem Soc 103:4239
14. Bonner FJ, Wolf R, Luisi PL (1980) J Solid-Phase Biochem 5:255
15. Olofsson G, Kizling J, Stenius P (1986) J Colloid Interface Sci 111:213
16. Smith RE, Luisi PL (1980) Helv Chim Acta 63:2302

Received September 30, 1986;
accepted November 7, 1986

Authors' address:

K. Holmberg
Berol Kemi AB
Box 851
S-44401 Stenungsund, Sweden

Progress in Colloid & Polymer Science   Progr Colloid & Polymer Sci 74:103–107 (1987)

Protein exchange reactions on solid surfaces studied with a wettability gradient method

H. Elwing, A. Askendal and I. Lundström

Laboratory of Applied Physics, Linköping University, Linköping, Sweden

Abstract: A surface wettability gradient method for the investigation of some quantitative aspects of protein adsorption on solid surfaces has been developed. The method has been used in studies of protein exchange reactions on silicon surfaces. The proteins used were human fibrinogen and γ-globulin and specific detection of these proteins was made using corresponding antibodies. By the gradient method, the protein exchange reactions could continuously be related to the wettability of the solid surface.

Key words: Protein exchange, wettability gradient method

Introduction

In most methods in the investigation of protein adsorption and desorption on solid surfaces, those with a constant chemical composition are used. If the influence of a chemical constituent is to be investigated it is often necessary to make several preparations of the surface, which is a laborious and expensive procedure with the strong possibility of a methodological error.

We have tested another approach, in which a surface constituent is deposited on the surface in a gradient over a distance [1]. The quantification of protein deposition on the gradient is made by the use of ellipsometry with sufficient lateral resolution. This procedure reduces the methodological error to a minimum.

In this communication we describe how the gradient method can be used for the investigation of protein exchange reactions continuously related to the solid surface wettability. A wettability gradient was obtained using a diffusion method to attach methyl groups onto spontaneously oxidized silicon surfaces.

Exchange reactions between human fibrinogen and γ-globulin adsorbed on the gradient were investigated. Specific detection of a given protein was made by using the corresponding antibody.

Material and methods

Proteins and antibodies

Human γ-globulin (HGG, 16.5% solution) and human fibrinogen (HFG, Grade L) were obtained from Kabi AB, Stockholm, Sweden. The proteins were dissolved in 0.1 M phosphate buffer containing 0.15 M NaCl at pH 7.3. Stock solutions of 20 mg protein/ml were prepared shortly before use. Swine-anti human IgG (anti-HGG) was obtained from Orion Diagnostica, Helsinki, Finland. Rabbit-anti HFG was obtained from Behring Werke AG, Marburg, F.R.G.

Preparation of gradient surfaces

The silicon used was boron doped, polished and 0.2 mm thick wafers (Wacker Chemie, Munich, F.R.G.). The wafers were cut into pieces of suitable size (10×25 mm) which were washed as described in [2]. The spontaneously formed silicon dioxide layer was about 1 nm thick as determined by ellipsometry. The pieces of silicon were placed in a cuvette filled with xylene (pro analysis, Merck). A 0.05% solution of $Cl_2(CH_3)_2Si$ (Merck) in trichloroethylene (pro analysis, Merck) was bedded under the xylene phase (Fig. 1). After 90 min of diffusion at room temperature the methylsilane/triphase was sucked from the lower part of the cuvette. The plates were then rinsed with ethanol and finally dried under a stream of nitrogen.

We have developed three methods for the determination of the wettability distribution along the gradient. One is a capillary rise method for the determination of advancing contact angles [1]. Another is to determine the amount of adsorbed fibrinogen on the gradient surface by the use of ellipsometry. This indirect method is based on the fact that there is a reproducible correlation between the surface wettability (determined as advancing contact angle with water) and the amount of fibrinogen adsorbed on the solid surface [3]. This "fibrinogen method" is useful for the determination of contact angles between 80° and 35° and was used in this investigation. A third method is to condense water on the gradient surface and observe the light scattering properties of the formed condensation drops on the surface [4, 5]. Light scattering is more intense at drops formed on the hydrophobic part. This qualitative but sensitive method can

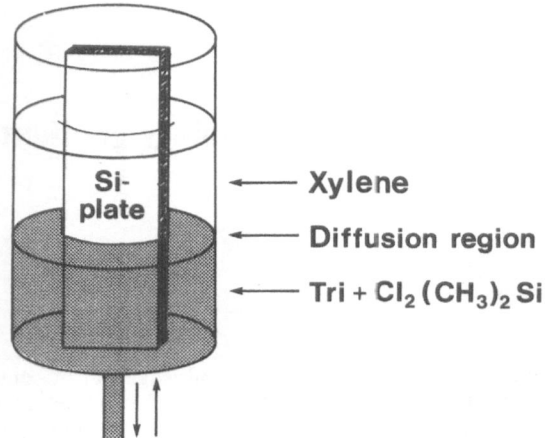

Fig. 1. Schematic illustration of the method used to prepare a surface gradient of methyl groups. Trichloroethylene (=tri) with 0.05% $Cl_2(CH_3)_2Si$ is bedded under the xylene solution, which has a low density. The methyl silane will diffuse to the xylene region and reacts with the solid surface. The process is stopped by removal of the solutions through the drain

be used to determine the position of the gradient and was also used in this investigation.

The advancing contact angle with water at the far hydrophobic and hydrophilic ends of the gradient plates was determined by an ordinary contact angle measurement method [6] and was found to be 85° and 16°, respectively.

The surface gradients were very stable. Immersion in 0.1 M buffer for 24 h with pH ranging from 2–8 did not change the surface wettability in a measureable way.

Protein exchange experiments

Silicon pieces with the gradient in surface wettability were incubated in either HGG or HFG (1 g/l) for 1 h at room temperature, followed by rinsing in distilled water. The pieces were then further incubated for 4 h in HFG or HGG (1 g/l) during gentle stirring at room temperature. Some of the pieces were also incubated for 30 min in anti-HGG or anti-HFG diluted to 1/25. All pieces were finally rinsed with distilled water and dried with N_2.

Ellipsometric quantification of adsorbed proteins

The dry gradient plates were mounted in an automatic ellipsometer (Auto Ell 2, Rudolph Research) equipped with a device for stepwise lateral scanning. The resolution of the lateral measurements was 0.635 mm and scanning was performed stepwise with 0.635 mm distance between each position.

The ellipsometer gives two angles Δ and Ψ which are related to the ratio between the complex reflection coefficients of p- and s-polarized light R_p and R_s, respectively

$$\varrho = \frac{R_p}{R_s} = \tan \Psi \, e^{i\Delta} \, .$$

From Δ and Ψ it is, in principle, possible to determine the refractive index and thickness of an organic layer adsorbed

on an optically characterized solid surface. The change in Ψ is, however, small for film thicknesses smaller than about 20 nm on a silicon surface. It is therefore possible to calculate both film thickness and refractive index independently, in the present experiments. Since we are measuring dried protein layers we assume that the protein has a density close to that of the bulk value ($\varrho = 1.4$ g/cm^3 was used in our calculations). Furthermore, we assume a fixed ratio between the molecular weight and molecular refractivity of the proteins (M/A = 4.1 g/cm^3 was used in our calculations). From this, we obtain an estimate of the refractive index of the protein, $n_p = 1.6$. With this n_p the average thickness of the protein layer can be estimated from the angle Δ. With the density ϱ_0 assumed to be above the adsorbed amount $\Gamma = \varrho_0$, d_p is then simply

$$\Gamma = 0.14 \, d_p$$

where d_p is in nm and Γ in µg/cm^2. This procedure is further described in ref [7] which also shows how the optical properties of the protein layer can be related to that of an equivalent silicon dioxide layer (with a refractive index = 1.46). This is convenient since the Auto Ell 2 is equipped with a program for calculating the thickness of silicon dioxide on silicon. Γ calculated in the described way is only an estimate of the adsorbed amount. The important information is, however, mainly in the relative magnitude of Γ in different cases. On the other hand, model calculations show that keeping M/A constant by using $\varrho_0 = 1.2$ or 1.5 g/cm^3, respectively, only changes the estimated Γ by about 10%.

The same density and ratio between the molecular weight and molecular refractivity were used for all proteins (HFG, HGG and antibodies). This may introduce an error in the comparison between adsorbed organic material in different cases, as shown in Figs. 2 and 3. Such detailed comparisons are, however, not necessary to draw the qualitative conclusions made later on in this paper.

Results

Adsorption of HGG on HFG preadsorbed gradient surfaces

The amount of adsorbed HFG ranged from about 0.75 µg/cm^2 on the hydrophobic part of the gradient to about 0.34 µg/cm^2 on the hydrophilic side (Fig. 2, line a). Gradient plates additionally incubated with HGG differed slightly from the former surface in that a smaller amount of deposited protein could be detected at the hydrophilic side of the gradient (Fig. 2, line b). Immunologically identifiable proteins were detected by the use of anti-HFG or -HGG. The additional amount of deposited antibodies is shown in Fig. 2, upper drawing. Incubation with anti-HFG resulted in a new deposition curve which immunospecifially indicates the presence of HFG (Fig. 2, line c). The adsorption of anti-HGG further indicated that HGG was identifiable on the hydrophilic but not on the hydrophobic side of the gradient (Fig. 2, line d). On gradient plates incubated with HFG only a small desorption at the hydrophilic side of the gradient was detected upon incubation with anti-HGG (Fig. 2, line e).

Adsorption of HFG on HGG preadsorbed gradient surfaces

The amount of adsorbed HGG differed only slightly between the hydrophilic and the hydrophobic side of

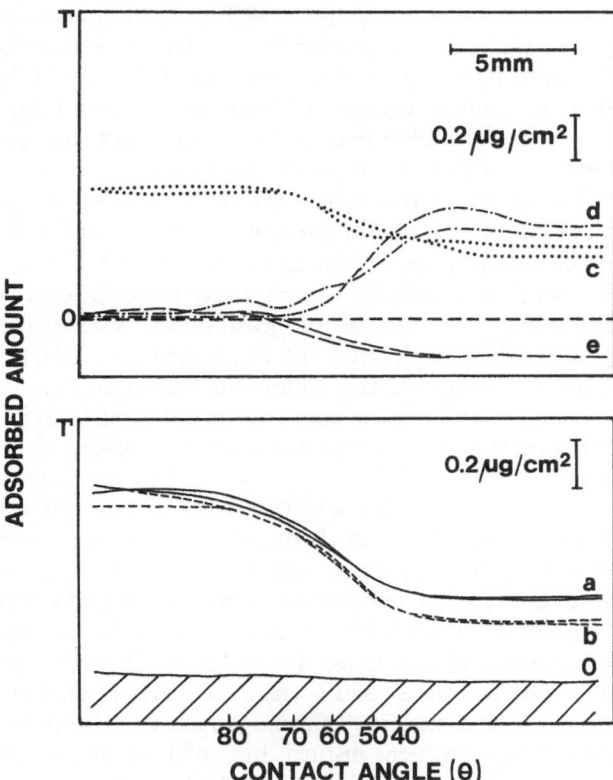

the gradient (Fig. 3, line a). After additional incubation with HFG, the adsorption pattern resembled that of HFG alone, with a larger amount of adsorbed HFG at the hydrophobic side of the gradient (Fig. 3, line b). Subsequent deposition af anti-HFG was large over the whole surface (Fig. 3, line c). Deposition of anti-HGG was small on the hydrophilic side and large on the hydrophobic side of the gradient (Fig. 2, line d). On gradient plates incubated with HGG followed by incubation in anti-HFG did not show any extra adsorption or desorption (Fig. 3, line e).

Discussion

The surface wettability gradient method appears to be well suited for the study of adsorption, desorption and exchange reactions on solid surfaces. Thus, the method has also been used in studies of detergent-induced desorption of proteins [1].

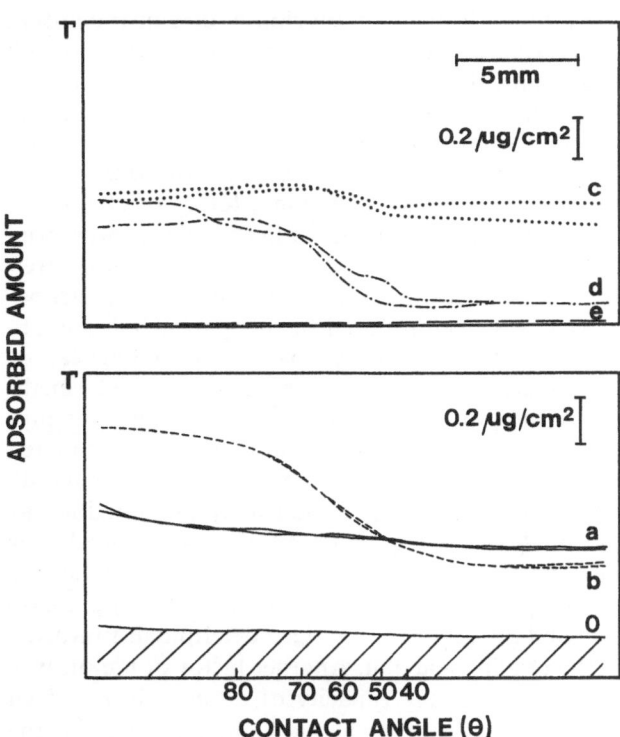

Fig. 2. Adsorption of HGG on gradient surfaces precoated with HGG. The adsorbed amount of protein was determined by ellipsometry as a function of distance along the gradient. The contact angles with water as determined by the adsorption of fibrinogen [3] are indicated on the abscissa. For clarity the results have been divided into two drawings. The bottom drawing shows: (a) adsorption of HFG followed by incubation with HGG; (b) adsorption of HFG followed by incubation with HGG; (0) marks the base line, i.e. the gradient surface without adsorbed proteins. The hatched area represents the oxide on the silicon surface plus a small contribution from the methyl groups. The top drawing shows the amount of deposited (c) anti-HFG on a surface precoated with HGG and then incubated with HGG (i.e. adsorption of anti-HFG on curve b); (d) anti-HGG on a surface precoated with a HFG and then incubated with HGG (i.e. adsorption of anti-HGG on curve b); (e) anti-HGG on a surface precoated with HFG. All experiments were performed in duplicate. In the present estimate, 0.2 µg/cm² corresponds to about 1.4 nm average thickness of the adsorbed layer of protein molecules. Further details are given in the text

Fig. 3. Adsorption of HFG on gradient surfaces precoated with HGG. The bottom drawing shows: (a) adsorption of HGG; (b) adsorption of HGG followed by incubation with HFG. The top drawing shows the amount of deposited: (c) anti-HFG on a surface precoated with HGG and then incubated with HGG (i.e. adsorption of anti-HGG on curve b); (e) anti-HFG on surface precoated with HGG (i.e. adsorption of anti-HFG on curve a). Further details are given in the text and in the caption of Fig. 2

The occurrence of protein exchange reactions is well documented [8–10]. Although the spontaneous desorption of most proteins is small in pure buffer, there are several experimental observations which indicate that protein molecules are exchanged on solid surface by protein from the solution. One explanation for the occurrence of exchange reactions has been discussed by Jennissen [11]. It is assumed that an adsorbed protein is attached to the surface by multiple binding sites. The binding sites are dynamic and disappear and rearrange constantly. It is unlikely that all binding sites of a protein molecule disappear at the same time. The rate of spontaneous desorption is therefore low. However, if other protein molecules are present in the solution, a single or a few binding sites can be replaced by those of a new molecule, which gradually occupies more binding sites and, with time, replaces the originally adsorbed molecule [11]. The details of the exchange reactions depend both on surface properties, type of adsorbed protein molecules, concentration and types of protein in solution and incubation time.

Although it has been previously demonstrated that exchange reactions take place more readily on hydrophilic than on hydrophobic surfaces, it is of considerable interest to have a method which directly relates the exchange reaction between different proteins to the surface properties. We have found that a very convenient way to obtain a lot of reliable information is to create a gradient in the surface properties of a well defined model substrate. The use of a wettability gradient created by hydroxyl- and methyl-groups was demonstrated in this paper, but it is obvious that the proposed preparation technique can be used to create other types of gradients also. The method of using specific antibodies to detect adsorbed protein antigens is convenient, but it should be pointed out that the method is indirect and precautions should be taken in the interpretation of results obtained by the method. The quantitative results obtained may thus be influenced by several factors, such as the amount of specific antibodies in the antibody preparation and the number of antigen determinants involved in the binding reaction. Another factor of importance is that the binding capacity of an adsorbed antigen does not always occur at the highest coverage of the antigen, due to stereospecific limitations [12]. There is also a possibility that the adsorbed protein may lose antigen determinants due to denaturation of the protein antigen at adsorption; new determinants may also appear in denatured molecules. Exchange of adsorbed proteins with antiserum proteins, especially at hydrophilic surfaces, must also be considered. Finally, if one molecule covers another on the surface, the covered molecule may be immunologically unidentifiable.

The model studies of human γ-globulin (MW = 160000) and human fibrinogen (MW = 340000) have again demonstrated that exchange reactions take place more readily on a hydrophilic surface. In addition the surface gradient method gives new analytical information. Thus, it is found that fibrinogen is partly exchanged by γ-globulin only on the hydrophilic side of the gradient, whereas on the hydrophobic side there appears to be no exchange at all(Fig. 2). It is also observed that fibrinogen adsorbs in a larger amount in the hydrophobic part of the gradient. It is very likely that the binding energy of fibrinogen is much larger on the hydrophobic part of the surface and thus exchange reactions are more difficult there.

The exchange reactions between fibrinogen and γ-globulin on γ-globulin precoated surfaces show an other pattern (Fig. 3). The test with the antibodies indicates that γ-globulin is completely removed on the hydrophilic side, but on the hydrophobic side of the gradient, both γ-globulin and fibrinogen are adsorbed after incubation of the γ-globulin coated surface in fibrinogen. In both the above described cases both incubation time and protein concentration influence the details of the results.

The preference for adsorption of fibrinogen on both hydrophilic and hydrophobic surfaces over various other proteins, notably albumin and γ-globulin, is a well documented effect [8–10] which is in accordance with the results obtained in this investigation. The exchange (or adsorption) preference of fibrinogen on a solid surface has been studied extensively, since adsorbed fibrinogen has an affinity for thrombocytes which, in turn, may play an important role in surface-induced clotting of blood. It has also been shown, however, that adsorbed fibrinogen on hydrophilic surfaces is rapidly exchanged by high molecular weight kininogen (HMWK, another clotting factor) from blood [13, 14]. The concentration of HMWK in the blood is at least eight times lower than fibrinogen. Despite this an almost total exchange of fibrinogen has been observed [15, 16]. The exchange reactions between fibrinogen and HMWK have also been documented by using the wettability gradient method [17].

We believe, that the model experiments presented above demonstrate how specially prepared gradient surfaces can be used, together with ellipsometry, to study interesting phenomena in connection with protein adsorption and interaction on solid surfaces. The usefulness of specific antibodies in this respect was also shown. Due to the low level of methodological error and high resolution, with respect to the wettability of the surface, we suggest that the gradient method offers new analytical possibilities in studies of different aspects of macromolecular interaction at solid-liquid interfaces.

Acknowledgement

Mr. Stefan Welin is acknowledged for the help with the ellipsometry experiments. This work was supported by grants from the National Swedish Board for Technical Development.

References

1. Elwing H, Welin S, Askendal A, Lundström I. J Colloid Interface Sci (in press)
2. Elwing H, Ivarsson B, Lundström I (1986) Eur J Biochem 156:359
3. Elwing H, Welin S, Askendal A, Lundström I (in press) J Colloid Interface Sci
4. Adams AL, Klings M, Fischer G, Vroman L (1973) J Immunol Methods 3:227
5. Elwing H, Nilsson LÅ, Ouchterlony Ö (1977) J Immunol Methods 17:131
6. Dahlgren C, Sundqvist T (1981) J Immunol Methods 40:171
7. Sternberg M, Nygren H (1983) J Phys C 10 Suppl 12:83
8. Brash JL, Samak QM (1978) J Colloid Interface Sci 65:495
9. Chan BM, Brash JL (1981) J Colloid Interface Sci 82:217
10. Brash JL, Uniyal S, Pusineri C, Schmitt A (1983) J Colloid Interface Sci 95:28
11. Jennissen HP (1981) Adv Enzyme Reg 19:377
12. Lundström I, Elwing H (1984) J Theor Biol 110:195
13. Vroman L, Adams AL, Fischer G, Munoz PC (1980) Blood 55:156
14. Vroman L, Adams AL (1986) J Colloid Interface Sci 111:391
15. Brash JL, ten Hove P (1984) Thromb Haemostas 51:326
16. Horbett TA (1984) Thromb Haemostas 51:174
17. Elwing H, Welin S, Askendahl A, Lundström I (in press) J Biomed Mat Res

Received September 4, 1986;
accepted December 28, 1986

Authors' address:

Hans Elwing
Laboratory of Applied Physics
Linköping University
S-58183 Linköping, Sweden

Progress in Colloid & Polymer Science   Progr Colloid & Polymer Sci 74:108–112 (1987)

Phase behaviour of binary and ternary non-aqueous aerosol OT systems

B. Bergenståhl, A. Jönsson, J. Sjöblom, P. Stenius and T. Wärnheim

Institute for Surface Chemistry, Stockholm, Sweden

Abstract: Binary and ternary phase diagrams have been determined for systems containing sodium di-2-ethylhexylsulfosuccinate (Aerosol OT); a polar solvent, formamide, methylformamide, dimethylformamide or ethanediol; with and without hydrocarbon, toluene or dodecane. In formamide, Aerosol OT forms a lamellar, a viscous isotropic (cubic) and a reverse hexagonal liquid crystalline phase. The swelling of the lamellar phase, as well as its temperature stability, is considerably reduced compared to the corresponding aqueous system. In the other solvents, only the reverse hexagonal phase forms. The ternary systems with toluene are generally dominated by an extensive solution region, i.e. a non-aqueous microemulsion, while with dodecane, the solution phase has a considerably reduced extension.

Key words: Aerosol OT, non-aqueous microemulsions, non-aqueous liquid crystals

Introduction

During the last few years, the interest in studying the aggregation of surfactants in non-aqueous but highly polar solvents has steadily increased. The formation and the properties of a lyotropic liquid crystalline phase of lecithin in 1,2-ethanediol, as well as in other diols, was studied in detail by Friberg [1]. The existence of a solvation force of similar magnitude as that in water has been demonstrated [2]. Aggregation in the solution phases has been investigated in different laboratories [3–5]. In addition to ethanediol, other strongly hydrogen bonding solvents (glycerol, formamide, to mention only some) were used. Of particular interest from a more fundamental point of view is the recent work on micelle formation by Evans et al., who used hydrazine als solvent [6, 7]. Ionic surfactants form micelles in hydrazine at concentrations similar to the aqueous systems. From measurements of the critical micelle concentration at different temperatures, Evans concluded that the micelle formation is due to the enthalpy term in the free energy balance, rather than to a gain in entropy of the type usually associated with the process in water [6, 7].

There are also different potential applications of non-aqueous surfactant systems where replacement of water in microemulsions by other polar solvents should be of great interest, including e.g. cleaning and using microemulsions as reaction media. However, only a few reports have been published in which solvents such as ethanediol, glycerol or formamide were used as replacements for water [8–10].

This paper, which is one part of a more extensive study of non-aqueous microemulsion systems, reports on the phase diagrams of binary and ternary systems containing Aerosol OT as the surfactant, different polar solvents and, in the ternary systems, hydrocarbon. For a zwitterionic amphiphile such as lecithin, it has been shown that the ability of the solvent to participate in hydrogen bonding is directly correlated to the swelling of the lamellar liquid crystalline phase [11]. No such systematic study has previously been reported for an ionic surfactant. A large part of the interest in the investigation clearly lies in the comparison between aqueous and non-aqueous systems. We will briefly try to rationalize the main features of the phase diagrams from a simple phenomenological point of view.

Experimental

Chemicals

Aerosol OT (Fluka) was used either as received or dissolved in methanol, which was evaporated, and the surfactant dried in vacuo to remove water. The water content was determined by Karl Fischer titration and by drying to constant weight, and was less than 1 weight %. Formamide (Aldrich 99%), N-methylformamide (Aldrich 99%), N,N-dimethylformamide (Aldrich 99%+), 1,2-ethanediol (Fluka 99%), toluene (Fluka 99%+) and dodecane (Fluka >98%) were used as received. The water content in these solvents was determined to be not more than 0.07, 0.15, 0.15, 0.10, 0.02 and 0.01 weight %, respectively, in solvent samples handled in a similar way to those used for the determinations of the phase diagrams. Comparisons of results obtained with dried and undried surfactant, respectively, indicated that small amounts of water do not have significant effects on the phase equilibria.

Phase diagrams

The boundaries of the solution phase were determined by titration of samples with hydrocarbon and with polar solvent in a thermostated bath at 25 °C. The phase transitions were detected by observation in plane polarized light or by visible effects on turbidity on passing a phase boundary.

Polarization microscopy

The liquid crystalline phases were identified between crossed polarizers by their appearance under the microscope. The same method was used to determine the temperature dependence of the phase equilibria in binary systems. The rate of change of temperature in these studies was 3 °C/min. A Reichert microscope equipped with a hot stage was used, usually at ×60 magnification.

Low angle X-ray diffraction

The diffraction patterns were recorded in a Guinier camera with a positon sensitive Tennelec PSD-100 detector (Tennelec Inc.) using Ni-filtered Cu-Kα radiation at 20 °C. The distance between the sample and the detector was kept a constant 46 cm. The detector was repeatedly calibrated with crystalline sodium octanoate with a repeat distance, d, of 23 Å.

Results and discussion

Binary systems

The phase diagrams for Aerosol OT with water and with formamide, respectively, are shown in Figs. 1a and 1b. Aerosol OT forms lyotropic liquid crystalline phases with formamide, including a lamellar phase. The formation of lyotropic liquid crystalline phases in a binary polar solvent/ionic surfactant system at room temperature have not, to our knowledge, been previously reported. However, comparison of the binary system Aerosol OT/formamide with the corresponding aqueous system indicates some differences. The formamide system forms a solution phase L, which is much more extensive than with water (Aerosol OT solubility in water at 20 °C is around 1.3 weight %). The swelling of the lamellar phase D is considerably

reduced in the formamide system; the D-phase is stable between 70 and 80 weight % surfactant, corresponding to a maximum molar ratio solvent/surfactant, 3/1. In water, the D-phase is stable up to 85 weight % solvent, i.e. the molar ratio solvent/surfactant, 140/1 [12]. A cubic [1] and a reverse hexagonal phase (F) exist within similar composition boundaries in both systems. Another noticeable feature is that the melting temperatures of all liquid crystalline phases are much lower in the non-aqueous system. In particular, the lamellar phase melts at 65 °C in formamide but is stable up to 150 °C in water.

From low-angle X-ray diffraction, the dimensions of the amphiphilic aggregates can be calculated, provided that the partial specific volumes of the com-

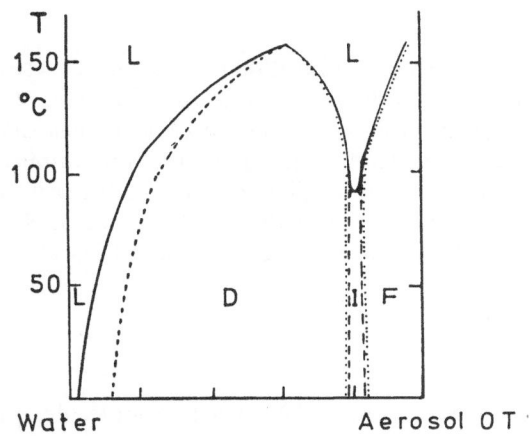

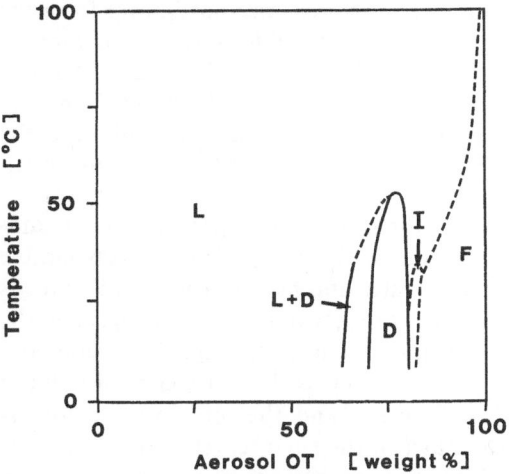

Fig. 1. Phase diagram for the systems: (a) Aerosol OT/water [12] and (b) Aerosol OT/formamide. (L) denotes isotropic solution phases (with or without subscript), (D) denotes a lamellar, (F) a reverse hexagonal and (I) a cubic liquid crystalline phase [15]

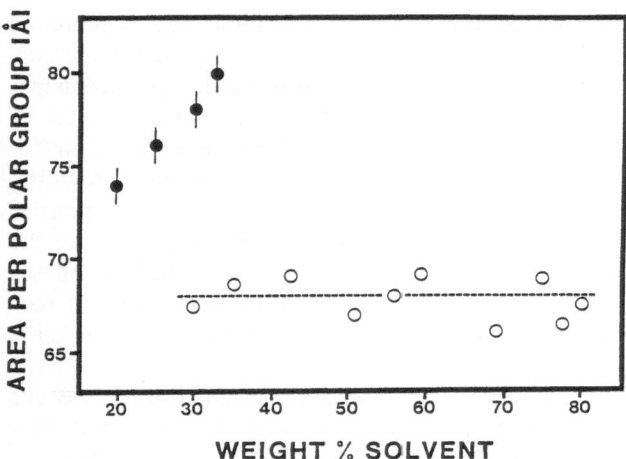

Fig. 2. Area per polar group at 20 °C for the D-phase in (○) Aerosol OT/water [13] and (●) Aerosol OT/ formamide

ponents at a given composition are known. If these are not available, volumetric data for the pure components or from similar systems may be used as an approximation instead. The approximation implies a small and normally negligible error, considering the error limits in the repeat distances. The area, S, per polar group can then be obtained according to

$$S = \frac{2V}{d\varphi}$$

where V is the volume of a surfactant molecule, d the fundamental repeat distance and φ the volume fraction of surfactant. The areas per headgroup from the lamellar phase in the formamide system are compared with the aqueous system [13] in Fig. 2. In the formamide system, the area increases from 73 to 80 Å^2 with increasing formamide content, while for the water system it is constant at 68 Å^2 per surfactant molecule.

The driving force for the aggregation of amphiphiles in water is the energetically unfavourable contact between water and hydrocarbon. This force is counterbalanced by repulsive interactions between the polar head groups. For a given aggregate geometry, the area per polar group is determined by a balance between this repulsion and the remaining hydrocarbon/water contact in the surface. At present, so little is known about non-aqueous systems that a detailed discussion of how this balance is changed is not possible. Some qualitative observations may, however, be made. The repulsion between head groups should be somewhat lower in formamide than in water because

of the higher dielectric constant (109.5 compared to 78.5 for water [14]). The expanded area per head group thus implies a lower interfacial tension at the aggregate surface in the formamide system. Also, in formamide and ethanediol, micelles may form but the critical micellar concentrations are considerably higher than in water [3, 4], showing that the driving force for aggregation is much lower. The increase in area at increasing solvent content is readily interpreted as an ongoing dissolution and destabilization of the lamellar phase. The general destabilization of ordered phases is also reflected in the lowered melting points of the liquid crystals that are formed (Fig. 1).

In the other solvents, methylformamide, dimethylformamide and 1,2-ethanediol, the reverse hexagonal phase formed by pure Aerosol OT is readily dissolved upon addition of solvent. For methylformamide and ethanediol, an isotropic solution phase occurs at more than 7 weight % solvent, while for dimethylformamide the figure is smaller. No liquid crystalline phase, other than the reverse hexagonal, occurs in these solvent systems.

Thus, the phase diagrams suggest that formamide behaves most similarly to water in the systems containing Aerosol OT and the various solvents. Only for formamide are the solvophobic effects sufficiently strong to give any appreciable aggregation into liquid crystalline structures.

Ternary systems

Phase diagrams for the ternary systems Aerosol · OT/toluene and formamide, methylformamide, dimethylformamide or ethanediol are shown in Fig. 3 a–d. The phase diagram for the corresponding aqueous system has not been determined. However, it should be very similar to the ternary phase diagram with p-xylene instead of toluene, shown in Fig. 3 e [13].

The reverse hexagonal F phase, which is predominant in the surfactant rich region of the aqueous system, has a reduced extension with formamide. It could be noted that slightly more than 30 weight% surfactant is needed to create a homogeneous solution of the otherwise immiscible formamide and toluene, i.e. a non-aqueous microemulsion (Fig. 3 a).

The disappearance of the miscibility gap between polar solvent/hydrocarbon upon addition of surfactant is in itself not necessarily indicative of surfactant action or aggregation. It has been known for a long time that the addition of a third component with comparable solubility in both the immiscible liquids will increase the mutual solubility of these liquids.

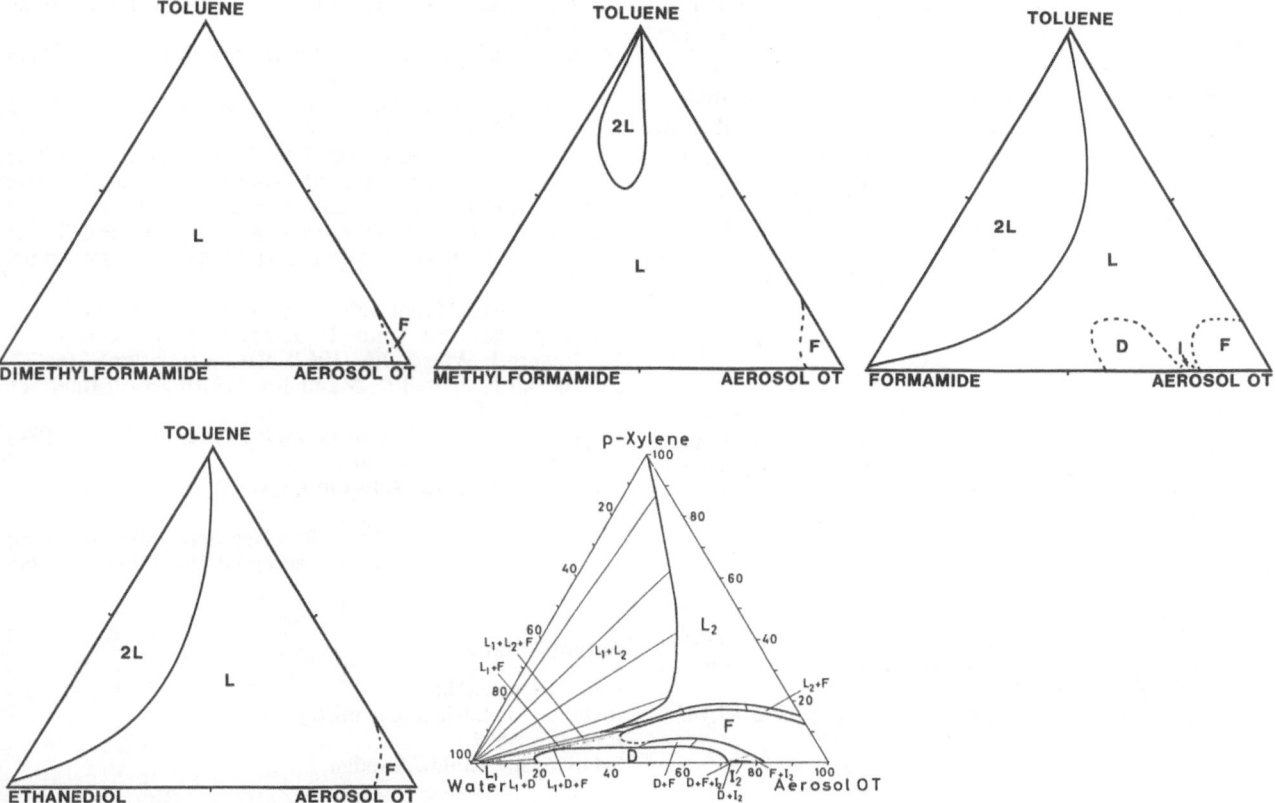

Fig. 3. Phase diagrams for the ternary systems: (a) Aerosol OT/formamide/toluene; (b) Aerosol OT/methylformamide/toluene; (c) Aerosol OT/dimethylformamide/toluene; (d) Aerosol OT/ethanediol/toluene at 25 °C and (e) Aerosol OT/water/p-xylene at 20 °C [13]. Notations as in Fig. 1, except that the D and F regions include two and three-phase regions at equilibrium with isotropic phases

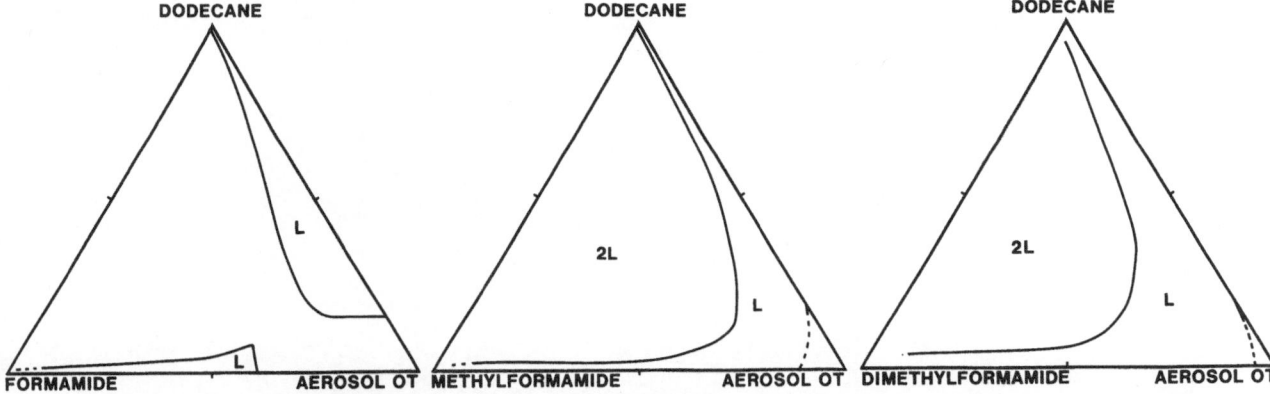

Fig. 4. Phase diagrams of the ternary systems: (a) Aerosol OT/formamide/dodecane; (b) Aerosol OT/methylformamide/dodecane; (c) Aerosol OT/dimethylformamide/dodecane, all at 25 °C. Only the extensions of the isotropic solution phases L are shown. The liquid crystalline phases have similar extensions as in the ternary systems in Fig. 3a–c

The ternary system with methylformamide (Fig. 3b) displays a very large solution phase. A curious feature is observed close to the toluene corner: a region in which two solution phases are at equilibrium, although methylformamide and toluene are complete- ly miscible at 25 °C. The dimethylformamide (Fig. 3c) is a single solution phase. In the ethanediol system (Fig. 3d) very small regions with liquid crystalline phases occur, while the miscibility gap is similar to that of the formamide system.

When the aromatic hydrocarbon is exchanged for an aliphatic hydrocarbon, the phase equilibria are changed considerably, as evidenced by Fig. 4a–c. Aerosol OT is, under these conditions, by no means optimal for the formation of extensive solution regions. This is made clear from the phase diagrams with dodecane, where the miscibility gaps are very large, in particular in the formamide system (Fig. 4a). For formamide, two separate solution phases occur due to the existence of liquid crystalline phases in the surfactant rich region.

Acknowledgments

This work was financially supported by the Research Council of the Swedish Board for Technical Development and is part of a joint program together with the Department of Physical Chemistry 1, University of Lund, for the characterization of non-aqueous microemulsions.

References

1. Moucharafieh N, Friberg SE (1979) Mol Cryst Liq Cryst 49:231; El-Nokaly MA, Ford LD, Friberg SE (1981) J Coll Interface Sci 84:228
2. Persson PKT, Bergenståhl B (1985) Biophys J 47:743
3. Ray A (1969) J Am Chem Soc 91:6511
4. Rico I, Lattes A (1986) J Phys Chem 90:5870
5. Wärnheim T, Henriksson U (1986) J Coll Interface Sci 112:66
6. Ramadan M, Evans DF, Lumry R (1983) J Phys Chem 87:4538
7. Ramadan M, Evans DF, Lumry R, Philson S (1985) J Phys Chem 89:3405
8. Rico I, Lattes A (1984) J Coll Interface Sci 102:285; Gautier M, Rico I, Ahmad-Zadek Samii A, de Savignac A, Lattes A (1986) J Coll Interface Sci 112:48
9. Friberg SE, Podzimek M (1984) Colloid Polym Sci 262:252; Friberg SE, Wohn CS (1985) Colloid Polym Sci 263:156
10. Magdassi S, Frank SG (1986) J Disp Sci Techn 7:345
11. Bergenståhl B, Stenius P (accepted) J Phys Chem
12. Rogers J, Winsor PA (1967) Nature (London) 216:477
13. Ekwall P, Fontell K, Mandell L (1970) J Coll Interface Sci 33:215
14. Handbook of Chemistry and Physics 64th ed (1984) CRC press, Cleveland
15. Ekwall P (1975) Adv Liq Cryst 1:1

Received September 30, 1986;
accepted December 11, 1986

Authors' address:

Björn Bergenståhl
Institute for Surface Chemistry
P. O. Box 5607
S-11486 Stockholm, Sweden

Progress in Colloid & Polymer Science

Progr Colloid & Polymer Sci 74:113–119 (1987)

Surface grafting of polyethyleneoxide optimized by means of ESCA

E. Kiss[1], C.-G. Gölander[2], and J. C. Eriksson[3]

[1] L. Eötvös University, Department of Colloid Science, Budapest, Hungary, [2] Institute for Surface Chemistry, Stockholm, Sweden, and [3] Department of Physical Chemistry, The Royal Institute of Technology, Stockholm, Sweden

Abstract: A hydrophilic, non-ionic surface was prepared by grafting poly-ethyleneoxide (PEO) onto polyethylene (PE) and mica substrates. Monofunctional PEO-aldehyde was coupled by means of reductive amination to primary and secondary amino groups of polyethyleneimine (PEI) adsorbed on oxidized polyethylene (PE) and on mica. Optimum values of the reaction parameters influencing the reductive amination in this heterogeneous system were sought in order to achieve complete coverage of the surface by covalently linked PEO chains. Carrying out the coupling reaction at 60 °C, pH 6 and $c_{K_2SO_4} = 11\%$ over 40 h resulted in a PEO surface characterized by a maximal ESCA $-C-O-/-CH_2-$ ratio and, simultaneously, complete wetting by water. A PEO surface with a high degree of coverage was also obtained by means of adsorption from solution of a PEO-amine compound on mica.

Key words: PEO surface grafting, ESCA characterization, polyethyleneoxide surface, aldehyde-amine coupling, minimal protein adsorption

Introduction

The purpose of surface modification by adsorption and/or chemical treatment of polymers is to achieve some particular new surface properties while retaining the bulk properties. Treatments yielding anti-fouling, anti-static, soil repellent, anti-bacterial surface properties etc., as well as biocompatibility, generally result in surfaces which are more or less hydrophilic. In order to reduce the adhesion of air-borne particles, a surface should be strongly hydrophilic while carrying a minimal number of ionic groups [1]. Exceptionally low degrees of protein adsorption have been observed on photopolymerized hydrophilic films which contain EO units and are attributed to the absence of ionic groups and the lack of strong hydrophobic interactions [2].

Our aim in this work was to develop a surface treatment applicable to various substrate materials resulting in a permanently bonded and densely packed monolayer of polyethyleneoxide (PEO). This could be achieved by making use of an aldehyde-amine coupling scheme and optimizing the reaction conditions, making use of ESCA to monitor the surface composition.

Experimental

Materials

A branched polyethyleneimine (PEI), Polymin SN, manufactured by BASF, F.R.G., was used at the surface treatment of the PE samples. PEI contains primary, secondary and tertiary amino groups in the approximate molar ratios 1 : 2 : 1. This PEI product was fractionated by ultrafiltration to obtain a molecular weight distribution between $10^5 - 10^6$. This Mw-fractionated PEI was also used in the mica experiments.

Polyethyleneoxide (PEO), or more precisely, polyethylene-glycolmonoethylether (Mw = 1900), having 43 ethyleneoxide units, was obtained from SIGMA, United States. PEO with one terminal aldehyde group (PEO-CHO) was prepared from PEO 1900 by partial oxidation in dimethylsulphoxide/acetic anhydride according to Harris et al. [3]. The product was purified by repeated precipitation in diethylether. A positive Schiff-test for aldehydes [4] and the appearance of an absorption maximum at 1735 cm^{-1} in the IR spectrum of the product indicated the success of the partial oxidation of PEO 1900.

Polyethyleneoxide-amine (PEO-NH$_2$) was prepared by further modification of PEO-CHO. Reductive amination of PEO-CHO with NH$_4$Cl and NaCNBH$_3$ was carried out as described by Harris et al. [3]. After purification of the product by repeated extraction with methylenechloride, the Schiff-test for aldehydes was negative while the test for amines with the Ehrlich-reagent [5] gave a positive result. All other chemicals used were of analytical grade.

Smooth and clean PE surfaces were prepared by slight melting of low density PE (Noax, Sweden) pressed between glass plates at 115 °C, followed by sonication in ethanol for 10 min.

The mica was delivered by SCIAMA, France. It was cleaved immediately before use in order to minimize carbonaceous surface contamination.

Sample preparations

Oxidation of the PE samples [6] was carried out in a concentrated H_2SO_4 solution of $KMnO_4$ ($c = 2$ g/l). The resulting "sulphated" PE surface ($PE\text{-}OSO_3H$) containing various polar groups such as $-OSO_3H$, $-OH$, $-COOH$ was rinsed thoroughly in distilled water.

Amino-functional surfaces were prepared by PEI adsorption from solution. Sulphated PE samples were exposed to 0.5% aqueous solution of PEI with pH adjusted to 3, 6 and 9 using borate and citrate buffers. After 10 min of adsorption the samples were rinsed with distilled water.

Clean mica surfaces were formed by cleaving and were immediately immersed in aqueous solutions of fractionated PEI and left for 15 min. The excess solution was removed by an N_2 jet.

In order to prepare PEO-surfaces, PEO-CHO was reacted in aqueous solution with the amino groups of PEI bonded to the PE or mica substrates. The coupling reaction can be considered as a reductive amination [7] of PEO-aldehyde. Primary and secondary amino groups of PEI bond reversibly to the PEO-aldehyde groups. Spontaneous dehydration in the presence of an acid catalyst gives imine or imminium salt intermediates which are subsequently reduced by $NaCNBH_3$ to PEO-amines. As this is a multi-step reaction, it is affected by the experimental conditions in a complex way.

The coupling reaction was carried out at different temperatures and salt concentrations. Reaction time and pH combinations were chosen according to a simple two-parameter experimental design scheme [8]. Our aim was to find the optimum conditions for realization of a high PEO-coverage.

PEO-containing surfaces were also prepared on mica substrates by adsorbing $PEO\text{-}NH_3^+$ ions from aqeous solution at pH = 6 for 2 h, followed by removal of excess solution with an N_2 jet.

Methods of analysis

A convenient colour reaction was used as a first qualitative test to select PEO-treated samples for further investigations. Polyethyleneoxides form orange-red precipitates in aqueous solution with the modified Dragendorff-reagent (tetraiodobismuthic acid + barium chloride) [9]. When used as a spot test this sensitive reaction is also capable of detecting PEO polymers which are attached to a substrate surface.

To characterize the wettability of the treated surfaces, contact angles of water droplets (2 µl) were measured at constant temperature and vapour pressure with a goniometer, model A100 from RAME-HART Inc., USA.

ESCA (Electron Spectroscopy for Chemical Analysis) spectra were recorded using a Leybold-Heraeus ESCA/ Auger spectrometer with an AlK_α anode ($h\nu = 1486.6$ eV) and a hemispherical electron energy analyzer. The operating conditions of the AlK_α source were fixed at 13 kV and 17 mA, and the sample surface was oriented 90 ° relative to the direction through the entrance of the analyzer. Detailed scans of C1s, N1s and O1s peaks were recorded over 10 min. Chemical shifts were determined relative to $E_b = 285.0$ eV for the $-CH_2-$ C1s peak. Deconvolution of the C1s peaks was made using a computerized Gaussian curve fitting program. Standard chemical shift values for carbon were taken

Table 1. ESCA characterization of PEO surfaces obtained by surface grafting under various conditions of PEO-CHO to sulphated PE substrates with PEI adsorbed at pH = 9

Coupling conditions				ESCA intensity ratios				Colour test for PEO
T/°C	t/h	pH	$c_{K_2SO_4}$/%	$\dfrac{O1s}{C1s}$	$\dfrac{S2p}{C1s}$	$\dfrac{N1s}{C1s}$	$\dfrac{-C-O-}{-CH_2-}$	
50	2	3.5	0.9% NaCl	0.326	0.036	0.147	1.6	−
50	2	3.5	11	0.349	0.029	0.107	1.8	−
60	3.5	6	11	0.335	0.020	0.098	1.7	−
80	69	7	9	0.431	0.011	0.031	3.6	+
60	25	6	11	0.450	0.013	0.057	2.6	+
60	25	7	11	0.430	0.005	0.026	2.9	+
60	28	5	11	0.444	0.015	0.039	3.0	+
*60	16	4	11	0.386	0.024	0.096	1.8	+
*60	16	6	11	0.462	0.018	0.067	2.7	+
*60	40	4	11	0.469	0.021	0.053	4.3	+
*60	40	6	11	0.461	0.008	0.018	6.5	+
60	16	7	11	0.436	0.005	0.047	3.6	+
60	40	7	11	0.419	0.006	0.034	3.8	+
60	40	8	11	0.418	0.004	0.035	3.6	+
Reference PEI surface without PEO-CHO				0.331	0.036	0.149	1.3	−

* Combination of experimental parameters, set according to experimental design.

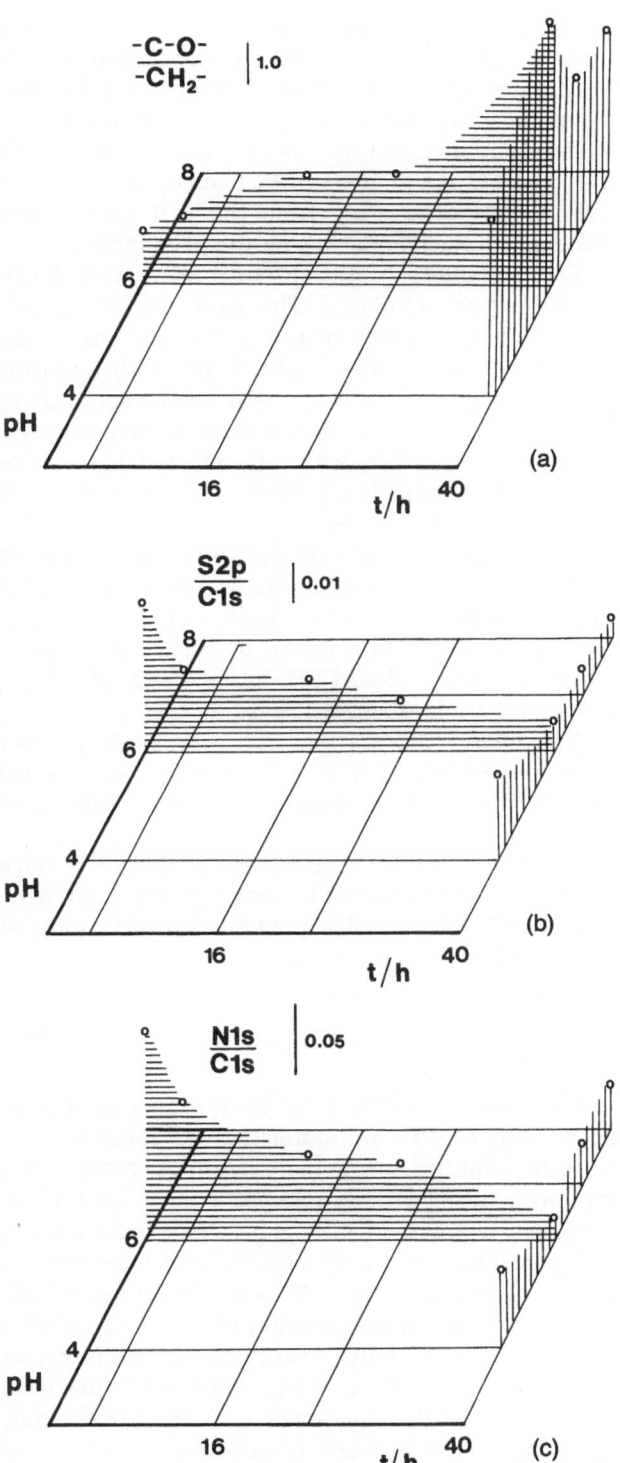

Fig. 1. Effect of two reaction parameters, pH and reaction time (*t* in hours) on the chemical composition of PEO surfaces obtained by coupling of PEO-CHO to PEI adsorbed on oxidized PE. ESCA intensity ratios of (a) −C−O− and −CH$_2$− peaks, (b) S2p and C1s peaks and (c) N1s and C1s peaks for *t* = 40 h and for pH = 6

from Gelius [10] and Scofield's relative cross sections [11] were used to calculate the stoichiometric ratios of elements: $\sigma(C1s)=1.00$; $\sigma(O1s)=2.93$; $\sigma(N1s)=1.80$; $\sigma(S2p)=1.68$; $\sigma(K2p)=3.97$.

Results

The main results concerning the influence of the reaction parameters on the coupling of PEO-CHO to amino-functional surfaces are summarized in Table 1. Reaction times in the range of 2 to 3.5 h proved insufficient to attach PEO chains onto the surface in any detectable amounts as judged from the colour test. For reaction times exceeding 16 h, the colour test for PEO gave a positive result.

The differences between the various PEO surfaces prepared can be studied by comparing the atomic composition data obtained from ESCA spectra. An oxidized PE surface with adsorbed PEI is used as a reference. Increases in the O1s/C1s and −C−O−/−CH$_2$− ratios indicate coupling of PEO to the amino-functional surface. Attenuation of the N1s and S2p signals originating from the inner layers simultaneously show an increased coverage by PEO. High −C−O−/−CH$_2$− ratios were, in most cases, only obtained after prolonged reaction times.

Addition of electrolyte to the PEO-CHO solution also has a significant influence on the outcome of the coupling reaction. Raising the salt concentration affected the atomic composition of the surface even for a reaction time of only about 2 h. A rather pronounced PEO coupling-enhancing effect was observed in the case of 11% K$_2$SO$_4$ at 60 °C.

In the pH region 4−6 and for a reaction time of 16−40 h the O1s/C1s and −C−O−/−CH$_2$− ratios increased with both parameters, while simultaneous decreases of the nitrogen and sulphur signals were noted. Further results at higher pH indicated less coverage of the surface by PEO, hence pH = 6 proved to be the optimum value for this two-phase coupling reaction. In Fig. 1 the ESCA intensity ratios of −C−O− and −CH$_2$− groups as well as the S2p/C1s and N1s/C1s ratios of PEO surfaces are displayed as functions of pH and reaction time at a constant value of one of the two parameters. There is evidently a tendency towards obtaining higher degrees of coverage at longer reaction times.

The effect of PEI adsorption on the coupling reaction with PEO-CHO was also investigated. PEI was adsorbed on the sulphated PE at pH 3, 6 and 9. The atomic compositions of the surfaces indicate that the amount of PEI adsorbed is about 30% larger at pH 9 than at pH 3 (Table 2). This is in line with the notable attenuation of the S2p/C1s ratio at high pH.

PEO surfaces with different degrees of EO-coverage were obtained by carrying out the coupling reaction on these PEI surfaces under the same (optimum) conditions. The degree of coverage by PEO was found to be

Table 2. Influence of the pH of PEI adsorption on the atomic composition of PEO surfaces

ESCA intensity ratio	pH = 3	pH = 6	pH = 9
a. PEI adsorption on oxidized PE			
$\dfrac{N1s}{C1s}$	0.111	0.136	0.149
$\dfrac{N^+}{\Sigma N}$	0.40	0.30	0.23
$\dfrac{S2p}{C1s}$	0.062	0.042	0.036
b. PEO−CHO coupling (T = 60 °C, t = 40 h, pH = 6, $c_{K_2SO_4}$ = 11%) to PEI adsorbed on oxidized PE			
$\dfrac{O1s}{C1s}$	–	0.436	0.461
$\dfrac{S2p}{C1s}$	0	0	0.008
$\dfrac{N1s}{C1s}$	0.019	0.021	0.018
$\dfrac{-C-O-}{-CH_2-}$	3.6	3.5	6.5

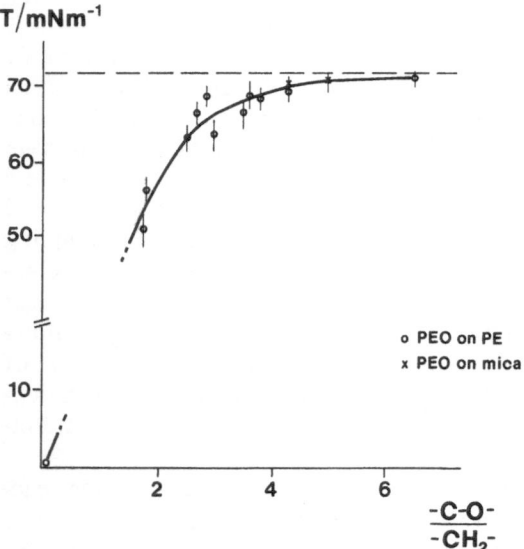

Fig. 2. Correlation between wettability (wetting tension: T) and $-C-O-/-CH_2-$ ESCA intensity ratios for PEO surfaces, $T = \gamma_{w,v} \cdot \cos \theta_w$, where $\gamma_{w,v}$ is the interfacial tension between water and saturated water vapour and θ_w is the contact angle of water droplet

much lower in the case of a low surface concentration of amino groups (pH 3, 6) than in the case of PEI adsorption at pH 9. The N1s/C1s intensity ratios after coupling are approximately the same (Table 2b). This is in accordance with the higher coverage of the surface by PEO at a high pH of PEI adsorption, since the N1s signals originating from the PEI surface also become stronger upon raising the pH (Table 2a).

The wettability of the different PEO-treated surfaces is presented in Fig. 2. The coverage of the surface by PEO is expressed as the ESCA intensity ratio of the $-C-O-$ and $-CH_2-$ peaks. Solids with a wetting tension above ~ 70 mN·m^{-1} may be characterized as water wetting surfaces. Such a water wetting property is achieved by surfaces with $-C-O-/-CH_2-$ ratios above ~ 4, independently of the substrate on which the PEO layer is formed.

Deconvoluted ESCA C1s signals for three different PEO surfaces are shown in Fig. 3. All C1s spectra are dominated by the $-C-O-$ peak, but the maximum surface coverage was achieved by surface grafting of PEO-CHO on sulphated PE. The $-C-O-/-CH_2-$ intensity ratio is 6.5 in that case.

A similar PEO surface but with slightly lower $-C-O-/-CH_2-$ intensity ratio (5.0) was obtained by applying the same surface modification treatment to a mica substrate.

The third spectrum (Fig. 3c) corresponds to a mica surface with lower but still rather high PEO coverage, $-C-O-/-CH_2- = 4.3$, prepared by adsorption of PEO-NH$_3^+$ ions.

Discussion

Amino-functional surfaces have been prepared by adsorption of PEI on sulphated PE substrates. In aqueous solutions, owing to its cationic character, PEI exhibits a strong affinity to anionic materials such as negatively charged organic solids. PEI when used in a highly protonated state adsorbs on oppositely charged interfaces in a flat configuration. The extent of adsorption of PEI increases with pH. This is attributed to the decrease in charge density of the PEI molecule and the higher surface charge density of substrates with weak acidic functional groups. Electrostatic repulsion effects between polycations adsorbed and those approaching the surface are diminished. Evidence for a flat adsorption pattern at low pH and for a "layered" structure at high pH was obtained by Larsson et al. [12] studying the angular dependence of the ESCA intensity ratio of neutral amine/protonated amine.

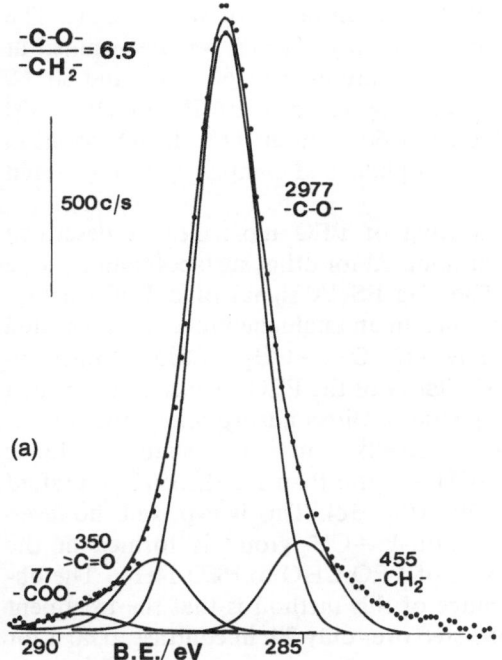

(a)

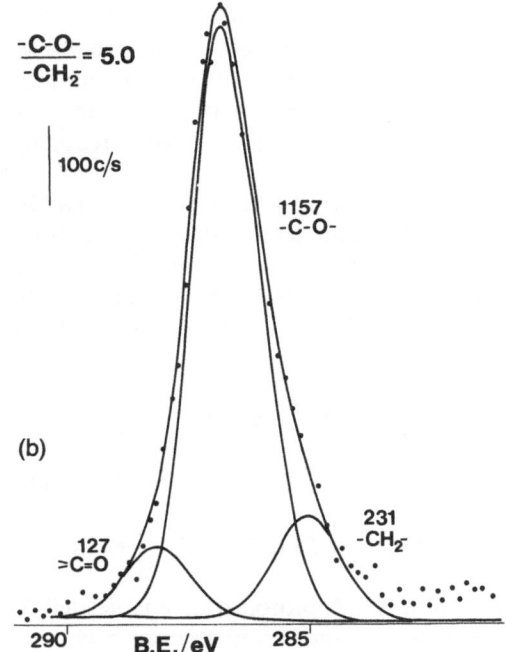

(b)

This is entirely consistent with our findings that coupling of PEO-CHO was most extensive with amino-functional surfaces obtained by adsorption of PEI at pH 9. This result is reasonable because of the relatively high surface density of amino groups at this pH, while the fraction of charged amino groups is retained high enough to attach PEI molecules firmly to the substrate surface.

The other reactant taking part in the coupling process, PEO-CHO, was prepared by partial oxidation of PEO under mild conditions resulting in a high yield [3]. Acetic anhydride activates the DMSO for reaction with the alcohol and the formation of the dimethyl-alcoholsulphonium salt intermediate is followed by an intermolecular hydrogen transfer giving the carbonyl product in this method [13, 14].

The coupling of PEO-CHO to an amino-functional surface is, of course a heterogeneous (two phase) reaction. Under these conditions it is not surprising that the usual reaction time of a few hours was found to be insufficient to generate the product in detectable amounts. The long reaction time required (order of 10 h) is probably a consequence of the fact that one of the reactants is anchored to a solid surface and diffusion is allowed only for the other reactant, thus lowering the probability of reactive molecular contact. An increase of the reaction time to above ~40 h did not result in any further increase in the coverage by PEO.

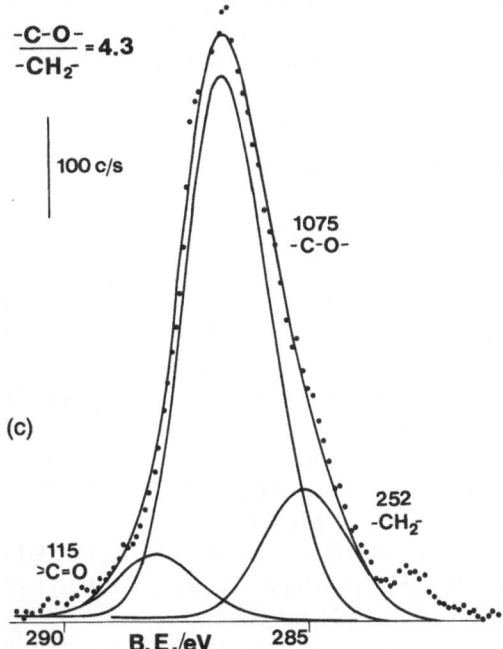

(c)

Fig. 3. Deconvoluted ESCA C1s signals for PEO layers formed on PE and mica: (a) PEO-CHO coupled to PEI adsorbed in oxidized PE, (b) PEO-CHO coupled to PEI adsorbed on mica, (c) adsorption of PEO-NH$_3^+$ onto mica. The intensity scale is in counts per second (c/s) and the binding energy (B.E.) scale is in electron volts (eV). Peak areas of each species and the intensity ratios of $-C-O-/-CH_2-$ are shown

The principle of the coupling reaction is reductive alkylation which is the term applied to the process of introducing alkyl groups (a PEO chain in our case) into ammonia or a primary or secondary amine by means of an aldehyde in the presence of a reducing agent.

The process of reductive alkylation includes a) addition of ammonia or amine to a carbonyl compound and b) reduction of the resulting compound or its dehydration product. The initial equilibrium reaction step forming an imine or imminium salt intermediate, i.e.

$$PEO\text{-}CH=O+NH_3 \rightleftarrows PEO\text{-}CH=NH+H_2O$$

$$PEO\text{-}CH=O+NH_2-R \rightleftarrows PEO\text{-}CH=N-R+H_2O$$

$$PEO\text{-}CH=O+NH\begin{subarray}{l}R\\R\end{subarray} \rightleftarrows PEO\text{-}CH=\overset{+}{N}\begin{subarray}{l}R\\R\end{subarray}+H_2O$$

is usually unfavourable. A proton source is needed to generate a positively charged condensation product.

Imines and imminium salts are readily reduced to amines by such varied reducing agents as sodium-borohydrate, hydrogen and Raney nickel, formic acid [7]. Simultaneously, the reduction of the aldehyde must be suppressed. Cyano-hydroborate is a remarkably selective reagent for this purpose [15]:

$$3\,PEO\text{-}CH=N-R+BH_3CH^-+H^++3\,H_2O$$
$$\rightarrow 3\,PEO\text{-}CH_2-NH-R+H_3BO_3+HCN$$

as the reduction of both aldehydes and imines is pH-dependent. Proper choice of pH thus provides a means to control the coupling reactions. pH = 6 was found to be the optimum for coupling of PEO-CHO to the amino groups of PEI indicating that reductive alkylation is favoured while reduction of PEO-CHO is still negligible at this pH.

Carrying out the coupling reaction at pH 7 or even 8 also resulted in PEO-covered surfaces but with lower degrees of coverage. At pH lower than 5 reductive alkylation is disfavoured on account of PEO-CHO reduction by $NaBH_3CN$ which acts as a nonselective reducing agent in this pH range.

To increase the surface coverage by PEO, favourable conditions for high PEO chain packing were chosen, so as to diminish the repulsion between the EO chains. The solubility of PEO in water decreases with increasing temperature [16] and this effect can be enhanced by addition of electrolytes. K_2SO_4 was chosen on the basis of experiments performed by Bailey Jr and Callard [17] to determine the cloud point of PEO as a function of the concentration of various salts. The optimum concentration of K_2SO_4 was determined at four different temperatures: 50°, 60°, 70° and 80°C. Finally, the coupling reaction of PEO-CHO 1900 when conducted at 60°C in an 11% K_2SO_4 solution resulted in a hydrophilic and completely PEO-covered surface.

Surface grafting of PEO molecules as described here might be adopted for other surfaces than PE (see e.g. [18]). The C1s ESCA signal of a PEO surface prepared on mica in an analogue process is presented in Fig. 3b. The $-C-O-/-CH_2-$ ratio obtained indicates the similarity of the PEO surface layer formed on the PE substrates. Direct adsorption of $PEO\text{-}NH_3^+$ onto mica results in a slightly lower $-C-O-/-CH_2-$ ratio than for the surface-grafted PEO-CHO film (Fig. 3c). This is expected, however, since an additional $-CH_2$-group is formed at the transformation of PEO-CHO to $PEO\text{-}NH_3^+$. The obvious advantage of this method is that the treatment occurs in one step over only 2 h in contrast to 40 h for the coupling reaction. The approximate surface concentrations of PEO 1900 were calculated from the C1s signal intensities according to Claesson et al. [19]. For PEO-lysine electrostatically adsorbed on mica, the area per molecule was found to be about 300 Å^2 [20]. In agreement with this, 320 Å^2/molecule was obtained for the adsorption of $PEO\text{-}NH_3^+$ on mica, while 190 Å^2/molecule was calculated for covalently bound PEO-CHO [18]. These values all imply that the PEO surface layer has a high water content [20].

Concluding remarks

Completely wettable non-ionic surfaces have been prepared by a surface modification treatment of PE and mica substrates. High degrees of PEO coverage were achieved by coupling of PEO-CHO at optimum reaction conditions to amino-functional surface layers adsorbed on the substrate.

The hydrophilic/hydrophobic nature of an EO unit appears to be rather well-balanced relative to the water-water interactions. This is evident, e.g. from the recent surface force experiments for hydrophobically adsorbed EO-surfactants carried out by Claesson et al. [21]. Very weak interaction forces were generally noted, except at short surface separations, where the EO-oligomers start to overlap, resulting in repulsion. Thus the driving force provided by the water for adsorption of a hydrophobic as well as of a hydrophilic/ionic solute is anticipated to be quite small, since little is to be gained free-energy-wise by displacing the water molecules attached to the EO units. Towards a more complete background of this nature we may eventually be able to rationalize the low degree of protein adsorption found at PEO-covered surfaces.

Acknowledgements

This work was supported in part by a grant from The Royal Swedish Academy of Engineering Sciences and The Hungarian Academy of Sciences. One of the authors (E. Kiss) gratefully acknowledges the opportunity for research work provided by the Institute for Surface Chemistry and the Department of Physical Chemistry at the Royal Institute of Technology, Sweden.

References

1. Larsson N, Eriksson JC (1982) In: Georges JM (ed) Microscopic Aspects of Adhesion and Lubrication. Elsevier Sci Publ Comp, Amsterdam, p 177−183
2. Gölander, CG, Jönsson S, Vladkova T, Eriksson JC, Stenius P (1986) Colloids Surf 21:149
3. Harris JM, Struck EC, Case MG, Paley MS, Yalpani M, van Alstine JM, Brooks DE (1984) J. Polym Sci Polym Chem Ed 22:341
4. Feigel F (1960) Spot Tests in Organic Analysis. Elsevier Sci Publ Comp, Amsterdam, p 675
5. Siggia H (1979) Quantitative Organic Analysis of Functional Groups, 4th ed. Wiley, New York
6. Eriksson, JC, Gölander CG, Baszkin A, Ter-Minassian Saraga L (1984) J Colloid Interface Sci 100:381
7. Allinger NL (1979) Organic Chemistry. Worth Publ, p 549
8. Box GEP, Behnken DW (1960) Technometrics 2:455
9. Brüger K (1963) Fresenius Z Anal. Chem. 196:251
10. Gelius, U., Héden PF, Hedman J, Lindberg B, Manne R, Nordling C, Siegbahn K (1970) Phys Scr 2:70
11. Scofield JH (1976) J Electron Spectrosc 8:129
12. Larsson N, Stenius P, Eriksson JC, Maripuu R, Lindberg B (1982) J Colloid Interface Sci 90:127
13. Albright J, Goldman L (1965) J Am Chem Soc 87:4214
14. Epstein WW, Sweat FW (1967) Chemical Reviews 67:247
15. Borch RF, Bernstein M, Durst HD (1971) J Am Chem Soc 93:2897
16. Florin E, Kjellander R, Eriksson JC (1984) J Chem Soc Faraday Trans 1, 80:2889
17. Bailey FE, Jr, Callard RW (1959) J Appl Polym Sci 1:56
18. Gölander CG, Kiss E. J Colloid Interface Sci (in press)
19. Claesson PM, Herder PC, Eriksson JC, Stenius P, Pashley RM (1986) J Colloid Interface Sci 109:31
20. Claesson PM, Gölander CG. J Colloid Interface Sci (in press)
21. Claesson, PM, Kjellander R, Stenius P, Christenson HK (1986) J Chem Soc Faraday Trans 1, 82:2753

Received December 12, 1986; accepted February 20, 1987

Authors' address:

Jan Christer Eriksson
Department of Physical Chemistry
The Royal Institute of Technology
S-10044 Stockholm, Sweden

Subject Index

Author Index